Research Reports ESPRIT

Project 369 · Vol. 1

Edited in cooperation with
the Commission of the European Communities

F. H. P. M. Habraken (Ed.)

LPCVD Silicon Nitride and Oxynitride Films

Material and Applications in Integrated Circuit Technology

 Springer-Verlag

Berlin Heidelberg New York London Paris
Tokyo Hong Kong Barcelona Budapest

Volume Editor

F. H. P. M. Habraken
Utrecht State University
Department of Atomic and Interface Physics
P. O. Box 80.000, 3508 TA Utrecht, The Netherlands

ESPRIT Project 369, Physical-Chemical Characterisation of Silicon Oxynitrides in Relation to Their Electrical Properties, belongs to Subprogramme 1, Advanced Microelectronics, of ESPRIT, the European Strategic Programme for Research and Development in Information Technology supported by the European Communities.

Project 369 was started in 1984 with the objective of establishing the relation between the physico-chemical and the electrical properties of silicon oxynitride films and the relation between the physico-chemical properties and the growth parameters. This research has been carried out in view of the likely applications of silicon oxynitrides in integrated circuit technology, for instance in MNOS- based non-volatile memories and in submicron MOS devices.

The project has identified a large number of material characteristics of silicon oxynitrides needed for improving the cost/performance ratio and the reliability of future generations of integrated circuits. Progress has been made in the basic understanding of the growth and characteristics of LPCVD and thermally grown silicon oxynitrides and several technologically important conclusions have been drawn regarding the film deposition process.

CR Subject Classification (1991): B.7.1

ISBN-13: 978-3-540-53954-4   e-ISBN-13: 978-3-642-76593-3
DOI: 10.1007/978-3-642-76593-3

Publication No. EUR 13343 EN of the
Commission of the European Communities,
Scientific and Technical Communication Unit,
Directorate-General Telecommunications, Information Industries and Innovation,
Luxembourg
Neither the Commission of the European Communities nor any person acting on behalf of the Commission is responsible for the use which might be made of the following information.

2145/3140 – 543210 – Printed on acid-free paper

PREFACE

The present book collects a broad overview of chemical and physical characteristics of silicon oxynitrides. Special emphasis is put on the way in which these properties influence the electrical characteristics and behaviour of this important material. The results presented here were obtained in an extended European research cooperation in the framework of ESPRIT Project 369 'Physical-chemical characterization of silicon oxynitrides in relation to their electrical properties', which ran from 1984 to 1988.

In this project two industrial laboratories (Philips Research Laboratories in Eindhoven, the Netherlands, and Matra Harris Semiconductors from Nantes, France) cooperated with various academic and government research laboratories (Harwell Laboratory in Great Britain, the Interuniversity Microelectronics Center (IMEC) in Leuven, Belgium, and the Faculty of Physics at the University of Utrecht in the Netherlands). The latter partner acted as prime contractor for the project.

General interest in silicon oxynitrides for applications in integrated circuit technology stems from the fact that proper choice of deposition conditions enables one to produce materials with properties which can be either oxide-like or nitride-like. Of course, in I.C. technology one would like to combine the good properties of the two materials, i.e. superior electrical properties of silicon oxide and good diffusion barrier behaviour of silicon nitride, to mention only a few, without paying for such an operation by obtaining all the less desirable properties in such a mixed material.

The present work turned out to be an exciting and challenging piece of materials research. In the project, which has concentrated on material obtained by low pressure chemical vapour deposition (LPCVD), much insight has been gained by measuring and comparing a very large number of samples. For instance, the all-important role of hydrogen for the structural, optical and electrical properties of silicon oxynitrides has been amply demonstrated. Nevertheless, some properties do remain elusive. Amongst these is the important problem of the exact nature of charge traps. In parallel with the research in this project, also applications have been considered. One successful application is the use of an oxynitride in the Local Oxidation of

Silicon (LOCOS) process. Another field in which silicon oxynitrides show promise is in thin gate oxides. In this application, use is made of thermally nitrided oxides.

The cooperation between the various partners in this project has been very effective. Not only has much insight in an interesting new material been gained, as stated above, but also, at least as important, experience has been gained in international cooperation, resulting in contacts which have proven to be important and fruitful, also beyond the duration of this project.

We acknowledge the financial support of the European Commission for this project and our EC project officers, Dr. B. O'Shea and Dr. M. Finetti, for their guidance, which was usually modest and in the background, but also very much to the point.

W.F. van der Weg, Project Leader
University of Utrecht

Table of Contents

Chapter 3
Oxidation of Low Pressure Chemical Vapour Deposited Silicon Oxynitride Films

A.E.T. Kuiper, M.F.C. Willemsen, J.M. L. Mulder, J.B. Oude Elferink,
R. Erens, F.H.P.M. Habraken and W.F. van der Weg

Chapter 4
Electrical Properties of LPCVD Silicon-Oxynitride Layers

M. Heyns, J. Remmerie, E. Dooms, H. Maes and R. De Keersmaecker

Chapter 5
On the Correlation between the Electrical and Physico-
Chemical Properties of LPCVD Silicon Oxynitride Films

F.H.P.M. Habraken, M. Heyns, H.E. Maes, R. de Keersmaecker,
A.E.T. Kuiper and W.F. van der Weg

Chapter 6
The Use of Oxynitride Layers in Non-volatile S-OxN-OS
(Silicon-Oxynitride-Oxide-Silicon) Memory Devices

H.E. Maes

Chapter 7
LPCVD Silicon Oxynitrides for LOCOS Isolation in CMOS Technology
J.L. Ledys

Chapter 1

CHARACTERIZATION OF LPCVD SILICON OXYNITRIDE FILMS

F.H.P.M. Habraken, J.B. Oude Elferink, W.M. Arnold Bik and W.F. van der Weg

Dept. of Atomic and Interface Physics, Utrecht State University

P.O. Box 80000, 3508 TA Utrecht, The Netherlands

and

A.E.T. Kuiper

Philips Research Labs

5600 JA Eindhoven, The Netherlands

and

J. Remmerie, H.E. Maes, M. Heyns and R.F. de Keersmaecker

Interuniversity Microelectronics Center

Kapeldreef 75, 3030 Leuven, Belgium

Abstract

The chemical composition of Low Pressure Chemical Vapour Deposited silicon oxynitride films is reviewed with emphasis on the hydrogen content and bonding configuration. Also effects of various anneal treatments are discussed. The data collected in this study can be interpreted assuming that in this Si-O-N system three basic structures exist: Si_3N_4, SiO_2 and Si_2N_2O.

I INTRODUCTION

Deposited silicon nitride films are used in a variety of applications in nowadays integrated circuit technology. In contrast to SiO_2 the nitride films have excellent diffusion limiting properties and therefore they exhibit a protective action against impurity diffusion and corrosion. Furthermore, they are not readily oxidized themselves, allowing, for instance, their application as oxidation mask in the LOCOS process (Local Oxidation Of Silicon). Another type of application lies in their dielectric properties for instance in metal-nitride-oxide-semiconductor (MNOS)-type memory devices or thin film transistors.

On the other hand SiO_2 films are preferably used in those circumstances where a large band gap is necessary for electrical insulation. The application of SiO_2 type of films as intermetal layer stems from this property. However, their poor diffusion limiting property makes the application of a nitride/oxide double layer necessary for the last purpose mentioned.

Investigations of deposited silicon oxynitrides are stimulated by the expectation that for application in i.c. technology the beneficial properties of silicon nitride and -oxide can be combined in one and the same material. The work is further encouraged by the fact that silicon oxynitrides can be grown via a simple extension of current Low Pressure Chemical Vapour Deposition (LPCVD) or Plasma Enhanced Chemical Vapour Deposition (PECVD) nitride or oxide processes. Therefore introduction of these materials is not hindered by major modifications of existing technologies.

This chapter deals with the physico-chemical characterization of LPCVD silicon oxynitride films. Also effects of anneal treatments will be discussed. It will be shown that the use of the high energy ion beam methods Rutherford Backscattering Spectrometry (RBS), Elastic Recoil Detection (ERD) and Nuclear Reaction Analysis (NRA) is very useful in this field. The information obtained from these techniques is complemented by (Fourier Transform) Infrared Spectroscopy (FTIR), a technique that may reveal the bonding configurations. Since hydrogen is supposed to play a role in the electrical performance of the material, emphasis is put on the hydrogen chemistry in the silicon oxynitrides.

II CHARACTERIZATION TECHNIQUES

An important factor in the characterization studies is the non-destructiveness of the techniques used. It emerged [1-3] that one must be very careful in the interpretation of measured spectra in electron spectroscopies like Auger and photoelectron spectroscopy because of artifacts due to the electron beam or to the ion beam used to sputter remove superficial oxide layers and to obtain depth profiles. For this reason, Wurzbach and Grunthaner applied chemical etching in conjunction with XPS to obtain a detailed compositional depth profile of MNOS structures [4].

It appeared that Rutherford Backscattering Spectrometry (RBS) applied under so called channeling conditions is suitable to determine the amounts of silicon, oxygen, nitrogen and chlorine in the silicon (oxy)nitride films [5,6,7]. Combining a channeling geometry with a low exit angle of the detected backscattered ions, a depth resolution of 1 nm appeared possible for very thin nitride films [7]. However, hydrogen cannot be detected using RBS. Total hydrogen concentrations may be determined with a relevant sensitivity using Elastic Recoil Detection (ERD), which is essentially the complementary technique of RBS [8]. When employing a primary beam of $\simeq$ 30 MeV Si ions ERD yields also accurate values for the amount of O and N in the oxynitride films. To obtain a depth distribution of H, Nuclear Reaction Analysis (NRA) is commonly used [8,9]. When the resonant reaction $^1H(^{15}N,\alpha\gamma)^{12}C$ is applied, a depth resolution of about 5 nm can be achieved in the surface region. Alternatively, H depth distributions with a relevant depth resolution may also be obtained by means of ERD employing low energy silicon ions [10] or 2 MeV He ions [11] as primary beam. Results obtained in this way are especially accurate if the analyses are performed under ultra high vacuum conditions [11].

Optical methods are highly rated on the scale of non-destructiveness. Infrared Spectroscopy has been used in the study of silicon (oxy)nitride films to measure Si-H and N-H bond concentrations [5,12-14] and to determine the O/N ratio in the material. Furthermore an attempt has been made to derive information about the structure of the material from IR spectra [14]. Ellipsometry is often used to determine the refractive index of the grown material, which is indicative for the composition. However, especially in the case of plasma grown oxynitrides, one must be careful using this procedure since it does not yield unambiguous results.

III RESULTS AND DISCUSSION

Bulk: Nitrogen, oxygen and chlorine

In this section results will be discussed which have been obtained in the study of LPCVD silicon oxynitride films about the incorporation of oxygen, nitrogen and chlorine. The films under consideration were grown from SiH_2Cl_2,

NH$_3$ and N$_2$O at temperatures around 800°C and at pressures of a few hundred mTorr [15].

Within a measurement accuracy of 5% stoichiometric Si$_3$N$_4$ films are grown when using a NH$_3$/SiH$_2$Cl$_2$ gas flow ratio of 2.5 or more. In these circumstances the growth rate is about 6 nm/min [6,16]. The Cl content is 0.06 at.% as deduced from RBS measurements [6].

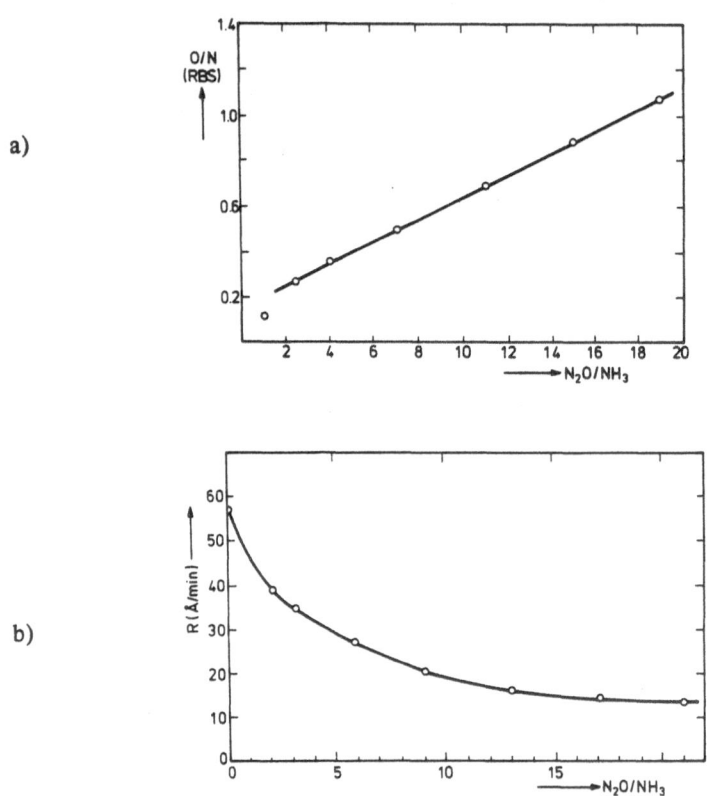

Fig. 1. O/N concentration ratio (a) and growth rate (b) as a function of the N$_2$O/NH$_3$ gas phase ratio [15].

The introduction of N$_2$O in the LPCVD reactor results in the incorporation of oxygen in the growing films [15]. Fig. 1a shows that the O/N concentration ratio in the oxynitride films increases linearly with the N$_2$O/NH$_3$ gas phase ratio; however, the N$_2$O/NH$_3$ gas flow ratio must have large values in order to incorporate oxygen in the films. At the same time the growth rate decreases with increasing N$_2$O/NH$_3$ ratio (fig. 1b). Apparently the reactivity of N$_2$O towards the growth surface is much lower than that of NH$_3$. The onset of growth of Si_3N_4 on Si, covered with a native oxide, is delayed [17]. It has been suggested that the native oxide has to be (partly) converted into an oxynitride via reaction with ammonia before the actual deposition can start at an appreciable rate [17]. This might indicate that not only the gas phase composition but also the composition of the growth surface is of importance for the deposition process.

In spite of the decreasing average reactivity of the gas phase molecules which do not contain silicon, and the declining growth rate, no significant amounts of excess silicon have been measured with RBS in the resulting silicon oxynitrides [15]. Excess silicon is defined as the amount of silicon that is in excess of the amount necessary to completely saturate the available O and N bonds.

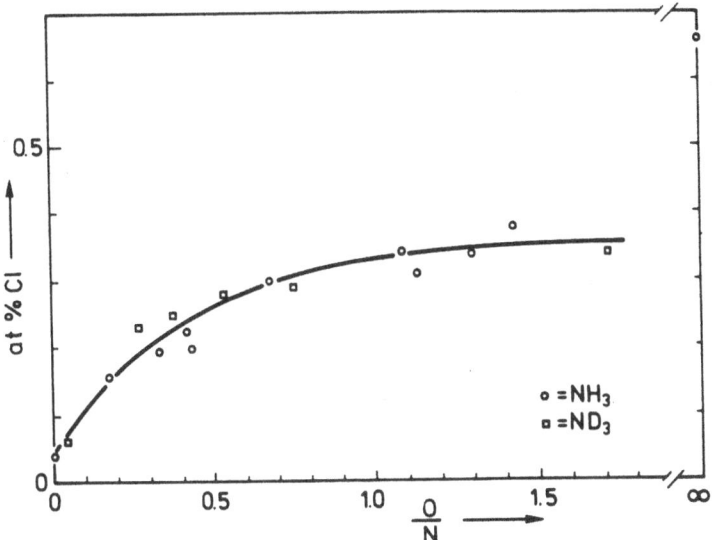

Fig. 2. *Cl concentration in LPCVD oxynitrides as a function of the O/(N+O) concentration ratio in the LPCVD oxynitride films [8].*

The Cl concentration in the oxynitrides increases with increasing O/(O+N) ratio from 0.06 at.% for the nitride to 0.65 at.% for the LPCVD oxide (fig. 2).

Bulk: hydrogen

The bulk H concentration in LPCVD silicon oxynitrides, as measured using

NRA, is given in fig. 3. For the lower oxygen contents [H] is independent of the O/(O+N) ratio [8] or increases slightly from 2.4 up to 3.5 at% at O/(O+N) ≈ 0.2-0.35 [18]. At larger O/(O+N) it decreases until at even larger oxygen concentrations it increases again [8]. Annealing at 900 and 1000°C in vacuum results in a decrease at all O/(O+N), but now a maximum in [H] occurs around O/(O+N) = 0.3-0.35. H present in the material around this composition is the most stable against heat treatment.

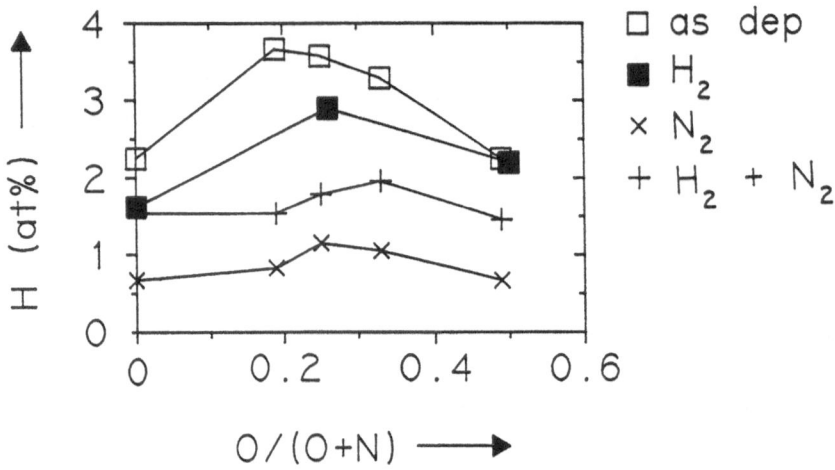

Fig 3. Bulk H concentrations in LPCVD oxynitrides as a function of O/(O+N) in the as deposited state and after annealing at 1000°C in ambients as indicated.

To obtain more information about the origin of H, oxynitrides from SiH$_2$Cl$_2$, N$_2$O and ND$_3$ were deposited. The amount of D incorporated relative to the amount of N has a maximum around O/(O+N)=0.3. After annealing this maximum is more pronounced, indicating that at that composition the deuterium is the most persistent in the film. Assuming that no isotope exchange occurs during deposition it was concluded that in the NH$_3$ grown films H is mainly bonded to N at not too large O/(O+N) ratios [8]. This has been corroborated by IR measurements [19]. For low O/(O+N) ratio there is quantitative agreement between the ERD deuterium data and the MIR IR measurements, but for O/(O+N)>0.3 deviations occur (fig. 4): in the IR study more N-H bonds are measured than D in the ERD study. An ERD study, in which 2 MeV He ions are employed and which has a better depth resolution, corroborate the 30 MeV Si ERD data [11].

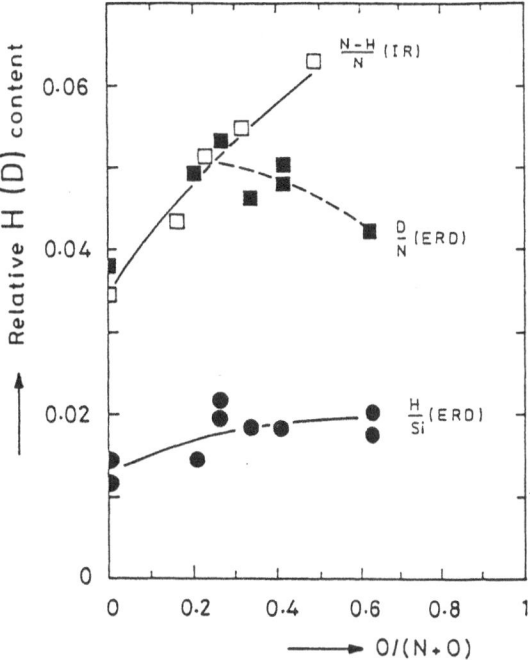

Fig. 4. *Amount of N-H (IR) and D (ERD) relative to the amount of N in LPCVD oxynitrides (from refs. 8 and 14).*

The Si-H bond concentration is below the detection limit of the IR technique. From an ERD analysis of the ND_3 deposited oxynitrides it is estimated that the amount of Si-H bonds is about 0.4 at.% for Si_3N_4 and does not vary much with increasing oxygen content at least up to $O/(O+N) = 0.5$ [8, 11]. There is an indication for a small maximum in the Si-H bond concentration at $O/(O+N) \approx 0.3$.

Except in a study of Peercy et al [20], concerning oxynitrides deposited in a Atmospheric Pressure CVD process from SiH_4, NH_3 and O_2, no indications for O-H groups in oxynitrides have been found.

At low $O/(O+N)$ ratios the IR and ERD N-H concentration measurements agree very well. This indicates that isotope exchange in the gas phase plays a minor role in the deposition of the deuterated films. Despite the decreasing N concentration in the films or the decreasing NH_3 flow in the reactor at increasing $O/(O+N)$, the amount of N-H is independent of $O/(O+N)$, at least for $O/(O+N)<0.3$. We conclude that a few percent of the NH_3 molecules do not completely decompose at the growth surface: they keep their last hydrogen atom. This fraction increases with increasing oxygen concentration for O/(O+N) < 0.3. Furthermore the N-H bonds become increasingly more stable against annealing with increasing oxygen content for O/(O+N)<0.3 [8]. These two effects are suggested to be interrelated.

Annealing

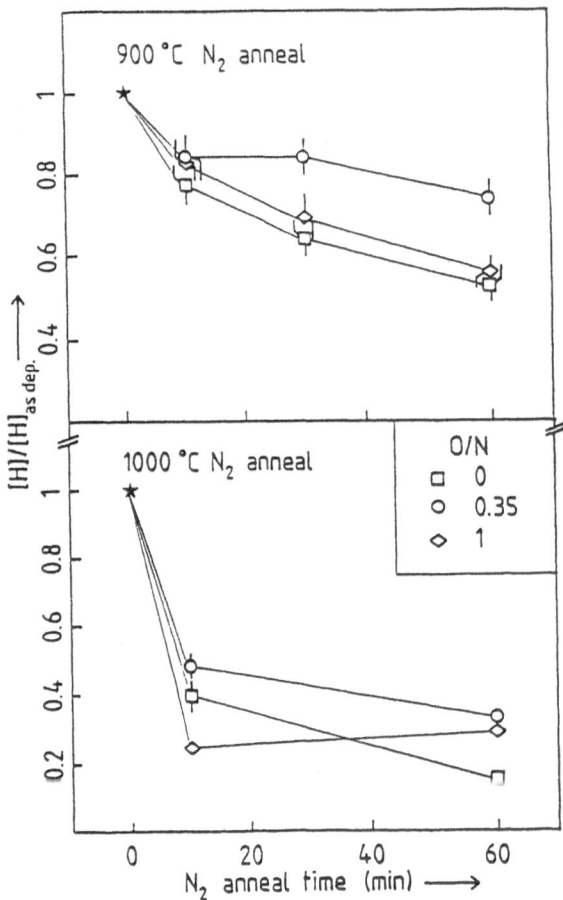

Fig. 5. The hydrogen concentration in LPCVD silicon oxynitride films as a function of anneal time at anneal temperatures of 900 and 1000°C in N_2.

It is known for some time that high temperature annealing of Si_3N_4 results in a degradation of the charge retention time in MNOS devices [21]. This has been related to outdiffusion of hydrogen during annealing, since subsequent annealing in hydrogen ambient restores (partly) the charge retention time [22-26]. During the hydrogen annealing step hydrogen is again introduced in the (oxy)nitrides [8,18,26]. So we studied the hydrogen outdiffusion and reintroduction in oxynitrides in more detail for three different O/(O+N) ratios (0, 0.26 and 0.5) [27].

Fig. 3 also shows the hydrogen contents in the oxynitride films of 70 nm thickness after various anneal treatments. A strong decrease of the hydrogen concentration is observed after 1 h $1000^{\circ}C$ N_2 annealing. The major part of the hydrogen loss occurs in the first ten minutes of the $1000^{\circ}C$ annealing (fig. 5). Subsequent annealing for 1 h at $1000^{\circ}C$ in H_2 results in an increase in H content in the oxynitrides. This uptake of hydrogen is already complete after 10 minutes of H_2 annealing (fig. 6). If the as deposited samples are annealed in H_2 without intermediate N_2 annealing, the hydrogen concentration is lower than in the as deposited films, although at O/N=1 this difference is very small. Also, the hydrogen content in the H_2 annealed films is larger than in the N_2+H_2 annealed films although for O/N=0 this difference is small.

It has been concluded from IR spectra that in the samples annealed in the

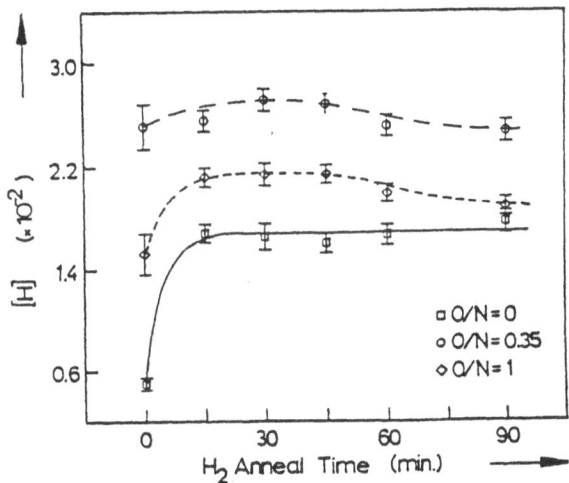

Fig. 6. The hydrogen concentration in LPCVD silicon oxynitrides after annealing at $1000^{\circ}C$ in H_2 as a function of the anneal time. O/N ratios for the films are indicated. Pretreatments: O/N=0: 60 min. $1000^{\circ}C$ N_2, O/N=0.35: 60 min. $900^{\circ}C$ N_2, O/N=1: 30 min $900^{\circ}C$ N_2.

hydrogen is bonded mainly to nitrogen; no Si-H or O-H bonds could be detected.

An explanation of these data necessarily requires an understanding of the mechanism of hydrogen escape during high temperature annealing. In the following a framework is developed within which some of the data can be understood.

From annealing studies of plasma grown (PECVD) silicon oxynitrides [28] it has been deduced that single bond breaking of N-H bonds does hardly occur at anneal temperatures below about 900°C. In PECVD oxynitrides containing both Si-H and N-H bonds hydrogen escapes already at anneal temperatures as low as 500°C. As evidenced by IR spectroscopy this escape of hydrogen is accompanied by formation of Si-N bonds, and by a decrease of the geometrical thickness of the films. So it is argued that the possibility for cross linking between Si and N opens a channel for hydrogen escape at relatively low temperatures.

In view of the high anneal temperatures of this study dangling bond formation as well as cross linking is expected to occur as a result of hydrogen escape from the films. It is not necessarily so that cross linking accompanies hydrogen escape as is suggested to occur in plasma grown oxynitrides; cross linking may also occur after removal of hydrogen. In the following we compare two models for the mechanism of hydrogen desorption. In one model, the hydrogen desorbs through diffusion of bonded hydrogen atoms to the surface from which it escapes into the ambient. In a second model hydrogen escapes via the formation of a fast diffusing H_2 molecule in the bulk of the film, for instance via cross linking. In both cases a certain mobility of bonded hydrogen is a prerequisite, because the concentration of hydrogen is such low that the average distance between two bonded hydrogen atoms is 10-15 Å.

To investigate the mobility and the desorption mechanism of hydrogen in Si_3N_4, a nitride layer was grown on silicon up to a thickness of 100 nm. The lower half of this film was deposited using ND_3 instead of NH_3; the upper part was deposited using NH_3 as reactant gas. The H and D ERD spectrum af the as deposited film is given in figure 7a. [29]. The two parts in the film are clearly distinguished in the H profile in the spectrum and in the upper part of the nitride film no D is observed. We have to keep in mind that the cross section for recoiling of D using Si as primary beam is 4 times lower than that for recoiling of H. After annealing of the sample for three minutes at 1000 °C in Ar, the H and D profiles as given in figure 7b have developed. Deuterium

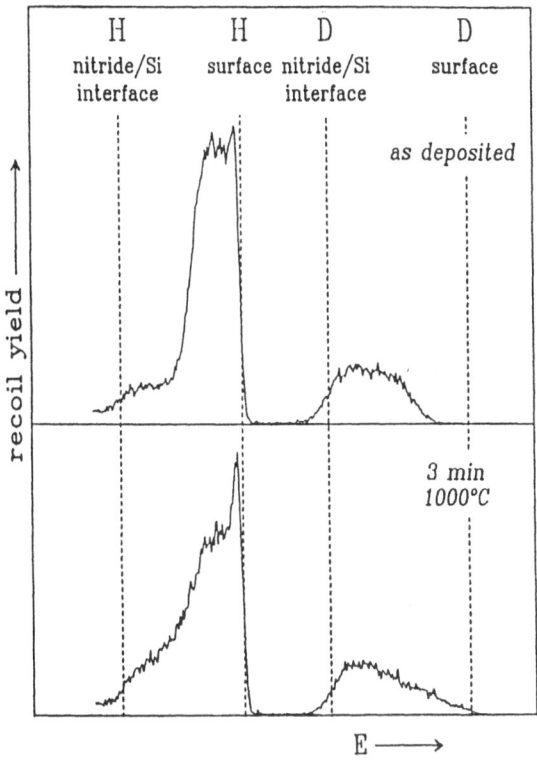

Fig. 7. ERD spectra of a silicon nitride film of which the lower half has been grown using ND₃ and the upper part using NH₃. The surface and interface positions are indicated. ERD measurement conditions: 10 MeV Si, take off angle 2⁰ and scattering angle 40⁰. The absorber foil thickness amounted to 4 µm. a) as deposited sample. b) sample annealed for 3 min. at 1000⁰C.

has clearly diffused to the outer surface of the film whereas part of the H has moved to greater depths. The total amount of D in this nitride double layer is given in fig. 8 as function of the anneal period at 1000 °C. It appears that D does escape measurably from the film after anneal times exceeding about 4 minutes. Inspection of the profiles such as given in fig. 7 learns that the anneal period of 4 minutes is sufficient for D to diffuse from the lower portion of the film to the surface. For comparison, figure 8 also shows the total amount of D in films in which the order of deposition of D was reversed: the layer deposited directly on the substrate, was grown using NH₃ and the upper portion of the film was deposited using ND₃. This implies that D

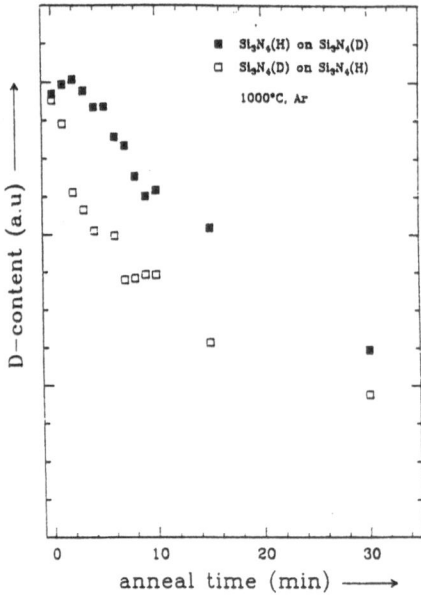

Fig. 8. Anneal time dependence of the total amount of D in nitride samples in which the ND_3 growth was performed onto NH_3 grown material and vice versa. The anneal temperature was $1000^\circ C$.

is present in the upper surface region of the nitride. Upon annealing of this sample at $1000^\circ C$ the total D content starts to decrease immediately. Results of Fourier transform infrared measurements of the N-D bond stretching mode intensity are consistent with the thought that all deuterium in the film is bonded. These experiments are interpreted as to represent strong evidence that the escape of hydrogen during high temperature treatment of silicon nitride proceeds through diffusion of bonded hydrogen to the surface and is effectuated via desorption from the surface region. This is in contrast with the model in which a fast moving H_2 molecule is formed in the bulk of the film. Since the hydrogen concentration profile over the depth of the film is flat [8,29], we assume that the rate limiting step is the desorption of hydrogen from the film surface.

As has been suggested to be the case in hydrogenated amorphous Si [30], hydrogen motion proceeds via hopping to neighbouring dangling bonds or through bond exchange processes. The escape of hydrogen from the film would then result in mainly N dangling bond defect sites. In this model cross linking is apparently not involved. But this does not imply that it does not occur. One argument for the occurrence of cross linking is that reintroduction of

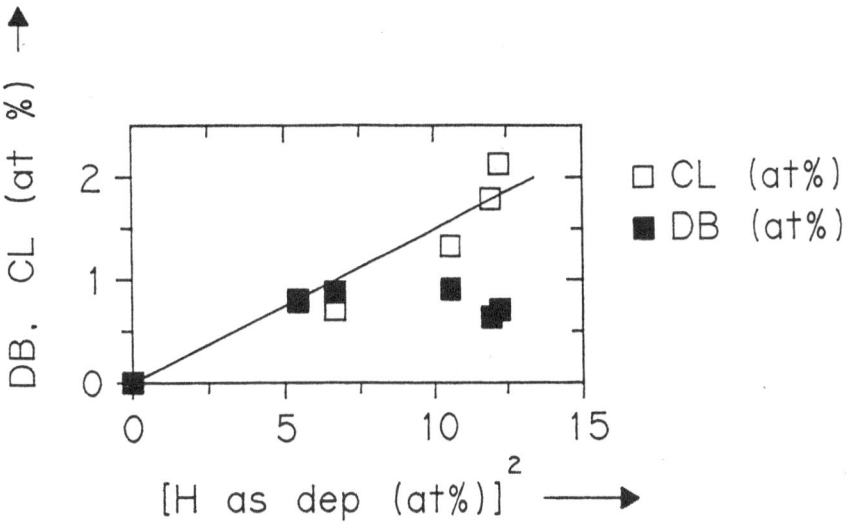

Fig. 9. The concentration of hydrogen desorbed from the oxynitride films via "cross linking" (CL) and "dangling bond formation" (DB) (see text) versus the square of the initial hydrogen concentration.

hydrogen in the oxynitride after a N_2 anneal through an H_2 anneal is never complete (fig. 3). From this observation we also tentatively conclude that H_2 is not able to break Si-N and Si-O bonds in significant amounts. Furthermore after H_2 annealing only N-H bonds have been observed. A further indication for the occurrence of cross linking may be found in the observation that the density of the material increases as a result of N_2 annealing [11].

If cross linking occurs, the amount of cross linked bonds may be estimated from the difference between the hydrogen concentration in the as deposited and the N_2+H_2 annealed samples. The concentration of dangling bonds arising in the N_2 anneal step, is estimated from the difference between the hydrogen concentration in the H_2 and the N_2+H_2 annealed films. In this estimations it is assumed that all present dangling bonds are passivated during H_2 annealing and at the same time that no bonds are broken in this heat treatment. The data of fig. 6 serve as a check that 60 minutes annealing certainly results in an equilibrium situation.

The concentration of dangling bonds and cross linked bonds, obtained via the above given procedure, are given in fig. 9. Since in first approximation the extent of cross linking is expected to increase quadratically with the initial hydrogen concentration, the data are plotted as a function of the square of the initial hydrogen concentration. Indeed, the concentration of cross linked bonds resulting from the 1 h 1000°C N_2 anneal increases about linear with the square of the initial hydrogen content. Consistently, the extrapolation to zero initial hydrogen concentration results in zero cross linking. The amount of dangling bonds does not vary significantly with the as deposited hydrogen concentration or, stated otherwise, with the O/N concentration ratio of the oxynitride.

Cross linking may occur between two Si-H bonds, a Si-H and a N-H bond and two N-H groups. Since the concentration of Si-H groups is only a few tenth of an atomic percent, we are tempted to conclude that mainly N-N bonds are formed during N_2 annealing. The possibility for pairing of nitrogen atoms to form N-N bonds, has been evidenced by Stein in his study of nitrogen implantation in silicon [31].

If we adopt the above given picture of the hydrogen behaviour during high temperature treatment, the results of the direct H_2 anneal experiments, i.e. the experiments in which the intermediate N_2 anneal has been omitted, may be interpreted as follows. At O/N=0 cross linking occurs to the same extent in a H_2 atmosphere compared to a N_2 atmosphere. This indicates that the cross linking rate in that material is comparable to the rate of formation of dangling bonds. In contrast, at O/N=1 the hydrogen concentration in the as deposited and H_2 annealed film is equal, which indicates that the cross linking rate in that material is slow compared to the rate of dangling bond formation.

Interface: oxygen and nitrogen

Probably the most detailed study of the LPCVD nitride/native oxide/silicon interface region has been published by Wurzbach and Grunthaner [4]. They reported that part of the native oxide still exists, while another part has been converted into an oxynitride. They also have reported that oxygen protrudes about 3-5 nm from the interface into the nitride at a low concentration, and that a thin layer (0.8 nm) with a significant amount of excess silicon exists about 1 nm away from the interface in the nitride. This excess silicon may be present in the form of Si-Si, Si-H bonds or in the form of Si dangling bonds. A conversion of the native oxide into an oxynitride and the existence of Si-Si bonds at the considered interface have also been observed by others [2,32]. However, a high resolution transmission electron microscopy examination of the Si_3N_4/SiO_2 interface showed that possible mixing of nitride and oxide, if occurring at all, must be limited to one or two atomic layers [33].

Single bond breaking of N-H bonds does not occur with an appreciable rate at temperatures below $900°C$ [28]. The deposition temperature of the LPCVD oxynitrides is only $800°C$. Therefore it is likely that deposition of the material takes place via cross linking of N-H and Si-H containing molecules or molecule fragments, impinging from the gas phase or adsorbed on the growth surface. A similar cross linking reaction may be operative to remove Si-Cl goups from the growth surface. During growth a large concentration of N-H, Si-H and/or Si-Cl groups will be present in the surface region. At the start of the deposition on the native oxide covered Si substrate this hydrogen and chlorine rich layer is not yet present, but has to be built up, for instance via a slight thermal nitridation of the native oxide before the "normal" (oxy)nitride deposition can take place. In this respect it is important to notice that thermal nitridation of SiO_2 gives rise to incorporation of hydrogen in the resulting nitroxide [34]. Within this picture the delay at the start of the deposition of silicon nitride [17] can be understood and it is also conceivable that the material in the interface region is different from that in the bulk.

Interface: hydrogen

Fig. 4 shows that for O/(O+N)>0.3 the ratio D/N as measured using ERD decreases with increasing oxygen content [8], whereas the concentration ratio N-H/N as measured with IR, increases [19]. Apparently, in this O/(O+N) region we are not allowed to equalize the deuterium concentration with the N-H bond concentration. This apparent discrepancy between the ERD and the IR results may result from the assumptions made to derive the relevant physico-chemical information. For the ERD data the critical assumption is that no isotope exchange occurs in the gas phase, on the growth surface or in the grown film. The interpretation of the IR measurements involves a constant absorption cross section for the N-H stretching vibration, which is beforehand not obvious

[33]. However, there is one parameter which is different in the IR and ERD study i.e. the thickness of the analysed films. The IR data are obtained from films with a thickness of 80 nm whereas the films used in the ERD study had a thickness of only 10-20 nm.

Furthermore, the anneal behaviour of hydrogen in the interface region appears to be different from that in the bulk as was deduced from a NRA study of H profiles in LPCVD silicon oxynitrides [18]. Fig. 10 shows the widths of the hydrogen concentration profiles in oxynitrides deposited on c-Si substrates, which were covered with a native oxide, and thin and thick thermal oxides after several subsequent anneal treatments. Within the depth resolution (~5 nm) the as deposited samples appear to have rather abrupt interfaces as far as hydrogen is concerned. However, the 1000°C N$_2$ annealed films show a large interface width (~20 nm) in the H profile in the case of O/(O+N)=0.5, irrespective of the substrate type.

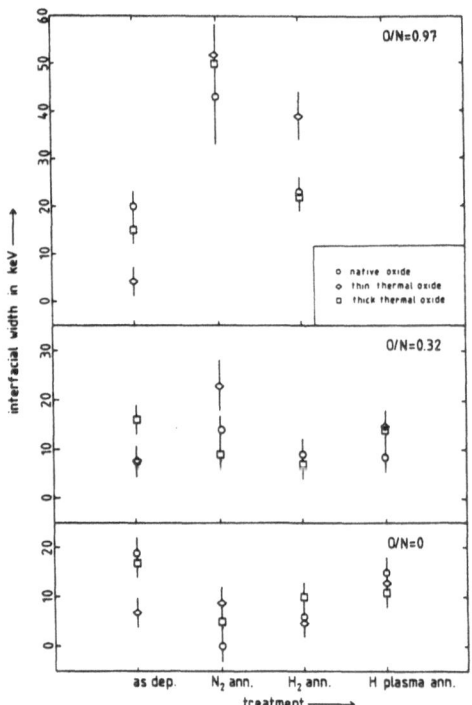

Fig. 10. The interfacial width of the hydrogen concentration at the interface of LPCVD SiO$_x$N$_y$ with c-Si covered with different SiO$_2$films. The O/N ratio and the treatments are indicated.

These observations motivated to investigate whether the hydrogen atoms are bonded differently in the interface region compared to the bulk of a film with O/(O+N)=0.58. In such a film of 100 nm initial thickness the N-H IR absorption peak area was measured in consecutively thin-etched layers. To this end this particular oxynitride was deposited on a MIR (Multiple Internal Reflection) crystal and after each removal of material an IR spectrum was recorded. In these experiments MIR crystals with 25 internal reflection were used. The layers and anneal treatments used in this experiment were identical to the ones that are used in the experiment for constructing the flat band voltage-thickness diagrams as described in chapter IV. This allows a direct

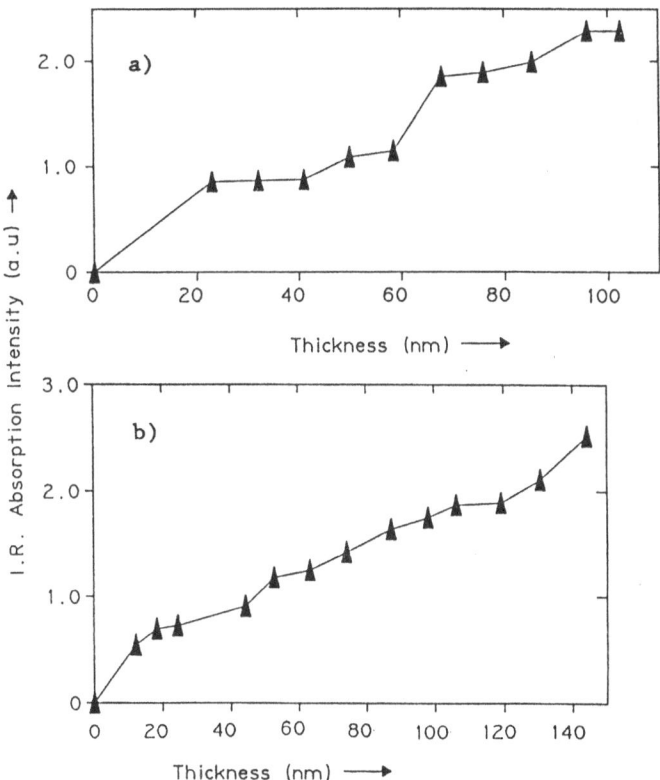

Fig. 11. *Infrared absorbed intensity in the region of the N-H stretching vibration as a function of the thickness of the oxynitride with O/(O+N)=0.58. a) single- and b) multi-step deposited sample.*

correlation between the H-related bonds and the charge distribution in the oxynitride layer.

In none of the spectra a Si-H peak was observed. This is caused by both the smaller extinction coefficient of the Si-H bond compared to the N-H bond and especially to the small concentration of Si-H bonds in the oxynitride.

The result of the measurement of the N-H absorption intensity for the as-deposited layer is shown in fig. 11a. The derivative of the curve through the measured data-points in fig. 11a is proportional to the local concentration of N-H bonds. Note that the interface position corresponds to zero thickness. The straight line that fits the data in fig. 11a indicates a uniform distribution of N-H bonds throughout the bulk of oxynitride. The positive intersection of this straight line with the y-axis is not significant within the accuracy of the measurement.

For 1000 $^{\circ}$C N_2+H_2 annealed and NH_3 annealed samples only an N-H peak is found but its area is too small to perform an in-depth profiling. The concentration of N-H bonds in the films nearest to 100 nm increases after applying an H_2 anneal after the N_2 anneal. An NH_3 anneal removes N-H bonds from the as-deposited sample but the N-H concentration after an NH_3 anneal is still larger than after an N_2 and an additional H_2 anneal.

MIR-measurements were also performed on multilayers formed by subsequent deposition of oxynitride layers to study the effect of a deposition on an underlying oxynitride film on the N-H distribution in an overlying film. The layer with an O/(O+N) ratio of 0.58 was deposited in three consecutive steps of approximately 50 nm each.

The measured area of the N-H peak versus the thickness of the oxynitride for the three-step deposited layer is shown in fig.11b. As for the one-step as-deposited layer a straight line can be drawn through the data indicating a uniform distribution of N-H bonds throughout the bulk of the oxynitride layer. Close to the interface a higher slope is found. A small accumulation of N-H bonds in this region may, therefore, be anticipated but is not resolved within the measurement accuracy. This was also found for the one-step as-deposited layers. Further, within the sensitivity of the measurement, no steps related to the three-step deposition can be observed in the data of fig. 11b. So, from the flatness of the NRA H profiles and the IR results of the thin-etched layers we conclude that, within experimental accuracy, the bonding configuration of hydrogen in as deposited oxynitride films does not depend on the distance to the interface.

Structure

Since the (LP)CVD oxynitrides are amorphous, the subject of the structure deals mainly with the question of the bonding configuration of the atoms. Consider a Si atom and neglect the presence of Si-Si bonds or Si-H bonds. That particular Si atom is coordinated by n (n=0,1,..,4) nitrogen and to m(=4-n) oxygen atoms. One may raise the question whether all values for n and m occur in an oxynitride with given O/(O+N). When for the majority of the Si atoms in a certain material holds that only the combinations n=4, m=0 and n=0, m=4 occur then a phase separation between oxide and nitride exist. On the other hand, if the probability for the occurrence of a certain configuration is solely determined by the relative number of N and O atoms then the materials can be conceived as a random bonding network.

It is commonly agreed that (LP)CVD silicon oxynitrides do not consist of a multiphase mixture of oxide and nitride. This was concluded from optical [36], XPS [37], Electron Energy Loss [1] and AES [2] analyses. However, this deduction does not imply that the oxynitrides are a random bonding network. Since the (LP)CVD of oxynitrides involves a decomposition reaction of reactant molecules which is supposed to proceed (partly) at the surface, the surface is also one of the reactants. Therefore the incorporation rate of a certain element may depend on the type of the incorporation site. In other words, chemistry may express its preference via the kinetics of the involved reactions, and hence the structure of the growing material is determined by the specific molecular reactions occurring at such charateristic sites.

In ref 20 it is mentioned that oxynitrides may be conceived as Si_3N_4 with O incorporated, or as SiO_2 in which N is inserted, depending on the O/(O+N) ratio of the material. This distinction in two basic structures to describe oxynitrides is corroborated by the observations that the refractive index, the growth rate and the etch rate change in a nontrivial way with the oxygen content [20,38]. The considered material was grown from SiH_4, NH_3 and O_2 at atmospheric pressure; especially the use of the very reactive O_2 may result in films which are different from the ones deposited using N_2O. However, in the study of Tombs et al [39] similar phenomena with respect to the etch rate and the refractive index have been found in N_2O-grown films.

A combined RBS/surface profile experiment by Kuiper et al [11] showed that the density (at/cm^3) of LPCVD oxynitrides relative to the density of a mixture of crystalline Si_3N_4 and SiO_2 of the same overall O/N ratio as in the LPCVD oxynitride, has a maximum around O/(O+N)=0.3-0.4. This may be an indication for a structural transition from nitride-like to oxide-like material at O/(O+N)=0.3-0.4. Note that this composition ratio corresponds to an overall stoichiometry near Si_2N_2O. So we tend to discern in this Si-O-N system three stable structures: Si_3N_4, SiO_2 and Si_2N_2O . The presence of SiO_2 and Si_3N_4 in this system is obvious; arguments for the presence of the Si_2N_2O structural units are found in certain material properties which vary with O/(O+N) but show a peculiar behaviour around O/(O+N) $\approx$ 0.3-0.4. Among these material properties are hydrogen chemistry, density [11], bulk charge etc (see also ch. V). The incorporation of oxygen in the nitride network adds some flexibility to the growing material because in the Si_2N_2O structure -SiN_3 units are bridged by twofold coordinated oxygen atoms instead of threefold coordinated nitrogen atoms.Experimentally, this increased flexibility and consequent decreased bond disorder may be deduced from the observation that the Si 2p and 2s photo-electron and especially the Si KLL Auger line starts to broaden with increasing O/(O+N) only at O/(O+N) $\approx$ 0.3 [35, ch. II]. This may be an indication that the increase in chemical disorder is compensated by the decrease in structural disorder. Furthermore, an increase in Si-N IR absorption frequency upon addition of small amounts of oxygen into the nitride has been interpreted as a decrease in bond disorder in the material [40]. In this respect it is important to notice that insertion of N-H groups in the nitride also adds flexibility to the nitride network since these groups are twofold coordinated by -SiN_3 units, similar to oxygen atoms. So $Si_2N_2(NH)$ is an isostructural analogue of Si_2N_2O. Therefore, the Si_2N_2O structure may be formed at an O/(O+N) ratio smaller than 0.33. With 2.5 at % H bonded to N, the Si_2N_2O structure may already be formed at O/(O+N)= 0.29. This is the value at which the N-D/N ratio shows a maximum and at which the D appeared to be the most persistent in the material against a 900°C anneal [8]. This observation emphasizes the structural component in the hydrogen chemistry.

Infrared Spectroscopy has frequently been used to study the bonding configurations in silicon nitride and oxide layers [12]. When analyzing oxynitride layers however, no separate bands appear in the wave number region for Si-O and Si-N stretching vibrations. One merely observes a broadened band at some intermediate frequency. Provided that a calibration has been performed the absorption peak position is a convenient and accurate measure for the oxynitride O/(O+N) composition (fig. 12).

The bending vibrations at 450 (Si-O) and 485 (Si-N) cm^{-1} have been reported to behave in a similar way. From the absence of separate absorption bands it was concluded that oxynitrides are no simple "mixture" or "codeposit" of two distinct phases SiO$_2$ and Si$_3$N$_4$ but can be described as "glassy atomic mixtures" or "polymers" of Si, N and O [14].

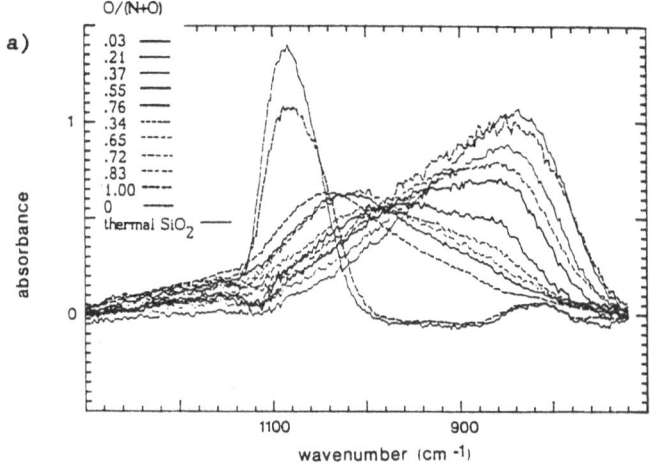

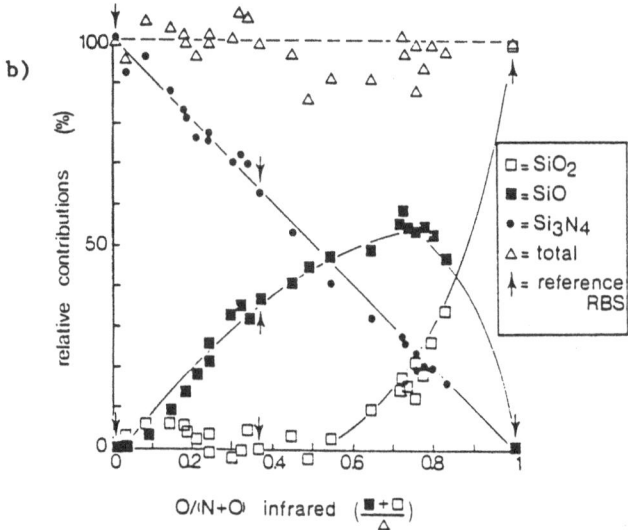

Fig. 12. a) Absorption bands in the 800-1200 cm^{-1} region of silicon oxynitride layers with different composition. b) The relative contributions of Si-N, Si-O and Si-O* as deduced from a PCA analysis of the infrared spectra of fig. 12a

A closer look at the spectra of fig. 12a reveals the apparent existence of (at least) three distinct features: a Si_3N_4 related contribution at ~840 cm^{-1}, a SiO_2-like absorption (asymmetric stretch) at ~1080 cm^{-1} which appears when the oxynitride composition closely approaches oxide stoichiometry, and an intermediate (broad) peak at ~1000 cm^{-1}. For very low oxygen contents no SiO_2-like contribution exists at all. The third (unknown) absorption feature has previously been supposed to be (mainly) oxygen-related and is, therefore, called SiO^* [14].

In order to obtain quantitative information on the contribution of these three species to the total absorption complex, a peak deconvolution technique is required. For this purpose a Principal Factor Analysis (PCA) was adapted. From the eigenvalue statistics it is confirmed that (at least) three independent spectral features are required in order to reproduce all spectra within experimental error. One has to be aware that the presence of three distinct absorption features does not necessarily imply the existence of three separate phases. The absorption frequency is merely an indication of the structural and electronic environment of the vibrating atoms.

The relative contributions of the Si-O, Si-N and $Si-O^*$ bonds as a function of the O/(O+N) ratio is represented in fig. 12b. The fact that PCA can not distinguish between more than three linearly independent components does not mean that no additional contributions are present. PCA only decides whether a satisfactory spectral reconstruction is possible within experimental noise limits. If additional information is taken into account, as the refractive index and relative permittivity for oxynitrides, one can try to decide on the total number of chemical contributions to the spectra on the basis of the non-random variation of the PCA-scalars with respect to the new information. It was found that at least two more separate components should be accounted for.

In the PCA theory proposed above, one supposed that the (three) contributions to the spectrum were linearly independent while the PCA analysis pointed out that the third component ($Si-O^*$) is in fact related to the Si-O bond which is already represented by another component. The theory is therefore only exact up to the first order. It would be more correct to replace the $Si-O^*$ contribution by terms which represent Si-O bonds interacting with several combinations of Si-O and Si-N environments. In this way the shape of the $Si-O^*$ contribution in the PCA analysis could be predicted.

Alternatively, the spectra in fig. 10a can be interpreted in terms of the "glassy atomic mixture". The Si-N stretch vibration in silicon nitride is located at 840 cm^{-1} and the Si-O stretch vibration in oxide at 1080 cm^{-1}. On traversing the oxynitride composition range from pure Si_3N_4 to pure SiO_2, this Si-N absorption will gradually decrease in intensity and the Si-O peak will grow correspondingly. Simultaneously, the spectral separation between these two modes will decrease simultaneously. This results from electronegativity effects on the bond strengths as pointed out by Lucovsky [41]. For example, when SiN_3O tetrahedra are present, the inserted oxygen will result in a slightly stronger Si-N bond and consequently in a shift to higher wavenumbers for the stretching mode. Correspondingly, the three N atoms will affect the Si-O stretch frequency to appear at a much lower wavenumber. In the oxynitride matrix all combinations of $SiN_{4-x}O_x$ tetrahedra will be present, their relative abundancies governed by the overall composition. Therefore the infrared peaks will be broadened and tend to overlap for intermediate compositions. In the PCA this overlap is accounted for by the so called SiO^* contribution.

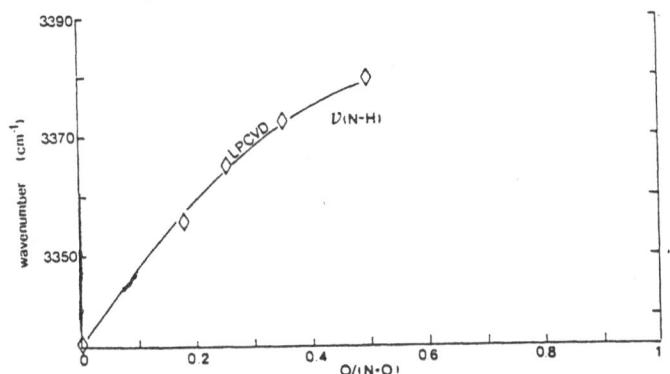

Fig. 13. The position of the N-H stretching vibration in LPCVD and PECVD oxynitrides and of Si-H in PECVD silicon oxynitrides.

For 0< O/(O+N) < 0.6 a continuous increase in the N-H stretching absorption frequency has been measured (fig. 13) [19]. This is expected when the considered N-H groups are to an increasing extent coordinated to groups which have a larger electronegativity than in the nitride (see above). This is the case when the oxygen content in oxynitrides is increased. This observation shows that at least in those regions where the N-H bonds are located, the material is not a multiphase mixture.

Stress

The stress in the as-deposited LPCVD layers is tensile for all measured oxynitride compositions and increases for a decreasing O/(O+N) ratio (fig. 14). An N_2 anneal at 1000°C for 1 hr lowers the stress for all O/(O+N) ratios and turns the stress even into compressive for an O/(O+N) ratio of 1. It was found that an additional H_2 anneal at 1000°C for 1 hr further decreases the

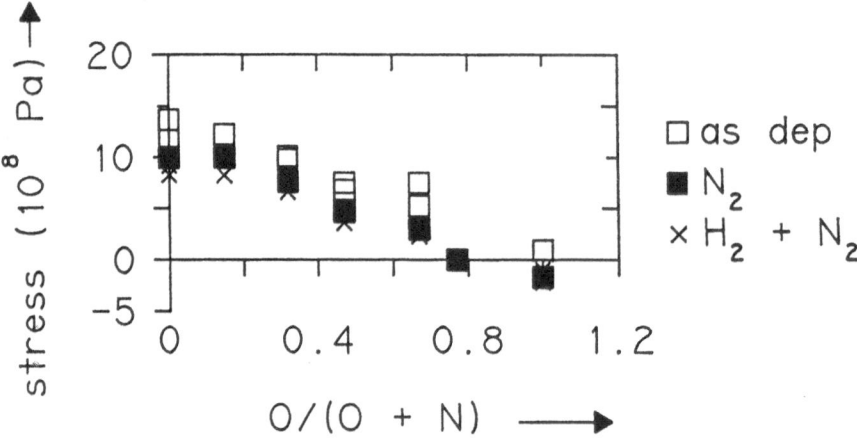

Fig. 14. *The stress in the LPCVD silicon oxynitride films as a function of O/(O+N) for various treatments.*

stress in the oxynitride layers.

The effect of an N_2 anneal or an H_2 anneal on the stress is a complex function of the different elements which generate the stress. A decrease (increase) of the hydrogen concentration is expected to increase (decrease) the stiffness of the layer and therefore the stress in the layer. On the other hand the possible occurrence of viscous flow for an anneal at 1000°C, or another structural rearrangement mechanism, may relieve part of the stress in the layer. This is evidenced by the lowering of the stress under a N_2 and an $N_2 + H_2$ anneal.

IV CONCLUDING REMARK

The results of the various analytical studies of LPCVD oxynitride films all point to the conclusion that this class of materials has an amorphous structure, which -depending on the overall composition- resembles that of Si_3N_4, Si_2N_2O or SiO_2. This structure is developed primarily in the process of deposition, but modifications may occur in postdeposition anneals, e.g. via cross linking reactions in the bulk of the films. Hydrogen plays an important role with respect to the structure: it is not only involved in densification via cross linking, but it also enlarges the existence range of the stable Si_2N_2O structure through the NH O compatibility. Although for obvious reasons it is difficult to provide solid evidence for this concept of three basic structures in silicon oxynitride, it allows for an elegant interpretation of published and collected data which would have remained troublesome otherwise.

REFERENCES

[1]. R. Hezel and N. Lieske, J. Appl. Phys. 51, 2566 (1980).

[2]. A. van Oostrom, L. Augustus, F.H.P.M. Habraken and A.E.T. Kuiper, J. Vac. Sci. & Technol. 20, 953 (1982).

[3]. J. Remmerie, Thesis University of Leuven, 1987. Ch. I.

[4]. J.A. Wurzbach and F.J. Grunthaner, J. Electrochem. Soc. 130, 691 (1983).

[5]. H.J. Stein, S.T. Picraux and P.H. Holloway, IEEE Trans. Electron Dev. ED-25, 1008 (1978).

[6]. F.H.P.M. Habraken, A.E.T. Kuiper, A. v. Oostrom, Y. **Tamminga** and J.B. Theeten, J. Appl. Phys. 53, 404 (1982).

[7]. Y. Tamminga, M.F.C. Willemsen, F.H.P.M. Habraken and A.E.T. Kuiper, Nucl. Instr. & Methods 200, 499 (1982).

[8]. F.H.P.M. Habraken, R.H.G. Tijhaar, W.F. van der Weg, A.E.T. Kuiper and M.F.C. Willemsen, J. Appl. Phys. 59, 447 (1986).

[9]. P.S. Peercy and H.J. Stein, *Proc. Symp. Silicon nitride thin insulating films*, eds. V. Kapoor and H.J. Stein ECS vol. 83-8 (1983) p. 3.

[10]. C.P.M. Dunselman, W.M. Arnold Bik, F.H.P.M. Habraken and W.F. van der Weg, MRS Characterization Bulletin, Ed. C.R. Helms, Vol. XII no. 6 p.35 (1987)

[11]. A.E.T. Kuiper, M.F.C. Willemsen and L.J. van IJzendoorn, Appl. Phys. Lett. 53, 2149 (1988),

[12]. H.J. Stein and H.A.R. Wegener, J. Electrochem. Soc. 124, 908 (1977).

[13]. W.A. Lanford and M.J. Rand, J. Appl. Phys. 49, 2473 (1978).

[14]. J. Remmerie, Thesis University of Leuven, 1987, ch. II.

[15]. A.E.T. Kuiper, S.W. Koo, F.H.P.M. Habraken and Y. Tamminga, J. Vac. Sci. & Technol. B1, 62 (1983).

[16]. P. Pan, J. Abernathey and C. Schaefer, J. Electronic Mat. 14, 617 (1985).

[17]. A.E.T. Kuiper, F.H.P.M. Habraken, A. van Oostrom and Y. Tamminga, Philips J. Res. 38, 1 (1983).

[18]. J.B. Oude Elferink, U.A. van der Heide, W.M. Arnold Bik, F.H.P.M. Habraken and W.F. van der Weg, Appl. Surf. Science 30, 197 (1987).

[19]. J. Remmerie, Thesis, University of Leuven, ch. IV.

[20]. P.S. Peercy, H.J. Stein, B.L. Doyle and V.A. Wells, *Proc. of the 7th International Conference on Chemical Vapor Deposition*, Los Angeles, edited by T.O. Sedgwick and H. Lydtin, Electrochemical Society, Pennington, NJ, 1979), Vol 79-3.

[21]. J.A. Topich and E.T. Yon, J. Electrochem. Soc. 123, 535 (1976).

[22]. H.E. Maes and G.L. Heyns, J. Appl. Phys. 51, 2706 (1980).

[23]. G.A. Schols, H.E. Maes and R.J. van Overstraeten, J. Appl. Phys. 51, 3194 (1980).

[24]. Y. Yatsuda, S. Minami, R. Kondo, T. Hagiwara and Y. Itoh, Jap. J. Appl. Phys. 19, 219 (1980).

[25]. J.A. Topich and R.A. Turi, Appl. Phys. Lett. 41, 641 (1982).

[26]. H.J. Stein, P.S. Peercy and R.J. Sokel, Thin Solid Films 101, 291 (1983).

[27]. J.L. Heideman, J.B. Oude Elferink, F.H.P.M. Habraken and W.F. van der Weg, to be published. J.B. Oude Elferink, Thesis, Utrecht State University, 1989.

[28]. C.M.M. Denisse, K.Z. Troost, F.H.P.M. Habraken, W.F. van der Weg and M. Hendriks, J. Appl. Phys. 60, 2543 (1986).

[29]. W.M. Arnold Bik, R.N.H. Linssen, F.H.P.M. Habraken and W.F. van der Weg, Applied Phys. Lett. june, 1990

[30]. R.A. Street, J. Kakalios, C.C. Tsai and T.M. Hayes, Phys Rev. B. 35, 1316 (1987).

[31]. H.J. Stein, in Proc. 13th *International Conference on Defects in Semiconductors* edited by L.C. Kimerling and J.M. Parsey (The Metallurgical Society of AIME, 1984), page 839.

[32]. R. Hezel and N. Lieske, *Proc. Insulating Films on Semiconductors*, Durham, 1979, Institute of Physics Conference Series no 450, eds. G.G. Roberts and M.J. Morant, p. 206.

[33]. Y. Kamigaki, S. Minami and T. Shimotsu, Appl. Phys. Lett. 53, 2629 (1988).

[34]. J.B. Oude Elferink, F.H.P.M. Habraken, W.. van der Weg, M. Heyns, E. Dooms and R.F. de Keersmaecker, Appl. Surface Science 39, 219 1989.

[35]. C.M.M. Denisse, J.F.M. Janssen, F.H.P.M. Habraken and W.F. van der Weg, Appl. Phys. Lett. 52, 1308 (1988).

[36]. H.R. Philipp, J. Electrochem. Soc. 120, 295 (1973).

[37]. J.C. Rivière, J.A.A. Crossley and B.A. Sexton, J. Appl. Phys. 64, 4585 (1988).

[38]. V.A. Wells and M.V. Hanson, *Proc. of the 7th International Conference on Chemical Vapor Deposition*, Los Angeles, edited by T.O. Sedgwick and H. Lydtin, Electrochemical Society, Pennington, NJ, 1979), Vol 79-3.

[39]. N.C. Tombs, F.A. Sewell and J.J. Comer, J. Electrochem. Soc. 116, 862 (1969)

[40]. H.J. Stein, *Proc. Vol. ECS meeting "Silicon nitride and silicon dioxide thin insulating films II"* Electrochemical Society Fall meeting, 1988, Chicago, Illinois.

[41]. G. Lucovsky, P.D. Richard, D.V. Tsu, S.Y. Lin and R.J. Markunas, J. Vac. Sci.& Technol. A 4, 681 (1986)

Chapter 2

SILICON OXYNITRIDE FILMS: ION BOMBARDMENT EFFECTS, DEPTH PROFILES, AND IONIC POLARISATION, STUDIED WITH THE AID OF THE AUGER PARAMETER

J.C. Rivière and J.A.A. Crossley,

Materials Development Division, Harwell Laboratory, U.K.

ABSTRACT

Thin films of Si_3N_4, SiO_2, and of silicon oxynitrides with compositions in the range $0.3 \leq O/N \leq 3.6$ were deposited on silicon substrates to thicknesses between 20 and 150 nm by Low Pressure Chemical Vapour Deposition (LPCVD). Auger Parameters of the films were measured during the study of ion bombardment effects and depth profiling, and as a function of the O/N ratios. The particular Parameter used was based on the Si 2s photo- electron, and Bremsstrahlung-excited Si KLL Auger, lines. When normalised to constant ion dose, no significant differences in the effects of bombardment with argon ions in the energy range 0.5 to 4.0 keV could be found, and the equilibrium O/N ratio measured by X-ray Photoelectron Spectroscopy (XPS) was close to that measured by Rutherford Back-Scattering (RBS) and Elastic Recoil Detection Analysis (ERDA). During depth profiling no reduction of the oxynitride to elemental silicon was observed, in contrast to previous work using the low-energy Si LVV Auger peak with electron excitation; it is therefore recommended that to perform depth profiling of these materials while retaining artefact-free compositional and chemical information, the high kinetic energy Si spectral features should be used. At the oxynitride/silicon interfaces in some of the films additional features were found corresponding to intermediate Si chemical states.

The widths of the Si 2s, 2p and KLL spectral features were measured at the stages in depth profiling at which equilibrium composition had been reached. In every case there was an increase in width as a function of oxygen fraction $O/(O+N)$, from Si_3N_4, through a maximum at about $O/(O+N) \approx 0.5$, followed by a decrease to SiO_2. The maximum increase for the KLL Auger peak was ≈ 0.9 eV, but for the 2s and 2p only ≈ 0.35 eV, and the peak broadening was uniformly distributed about the peak centroid. The broadening is attributed to the appearance of a number of additional chemical states of Si too closely spaced to be resolvable, and probably arising from a defective film structure.

The Auger Parameters of the various films, measured also once equilibrium composition had been reached during profiling, showed what appeared to be a monotonic dependence on $O/(O+N)$, thus supporting the random bonding model in which nitrogen atoms are substituted randomly by oxygen in the SiN_zO_{4-z} (z = 0, 1, 2, 3, 4) tetrahedra. However, the Auger Parameter vs. composition plot was displaced slightly from the theoretical one in the direction of higher Parameter than expected.

Reprinted in part from the Jornal of Applied Physics 64 4585 (1988)

with permissionn. © American Institute of Physics

1. INTRODUCTION

In Very Large Scale Integrated circuit (VLSI) technology
non-volatile memory devices are currently fabricated from either
metal-nitride-oxide-silicon (MNOS) or silicon- nitride-oxide-silicon
(SNOS) structures, in which the charge trapping characteristics are
determined by the dielectric properties of the silicon nitride thin
film, and by the electrical conditions at its interfaces. Such films
are also used as diffusion barriers and as oxidation masks (e.g. in the
so-called Local Oxidation of Silicon (LOCOS) technique), and for these
purposes are usually prepared by Low Pressure Chemical Vapour Deposition
(LPCVD). If, on the other hand, silicon nitrides are to be employed as
intermetallisation layers, as passivation caps, or as gates in FET's,
then they are prepared by the low temperature Plasma Enhanced Chemical
Vapour Deposition (PECVD) method.

For long-term stability and reproducible operation of a device such
as an MNOS it is necessary that several conditions be fulfilled, viz.
good chemical stability (i.e. resistance to oxidation), low interface
state density so that charge transport is minimised and charge trapping
optimised, and low diffusivity of impurities. As mentioned above, some
of these requirements can be met by using a nitride-oxide sandwich, but
it has been suggested[1] that it would be advantageous to replace the
double layer by a single film of silicon oxynitride SiO_xN_y, where the
ratio x/y is chosen to suit the particular application. Although such
films can be prepared by thermal[2] or by plasma[3,4] nitridation of silicon
dioxide, the nitrogen content is not homogeneous through the film;
nitrogen piles up at the surface and interface regions, and
incorporation in the bulk of the film is limited by diffusion. A much
better method[1], and one that is very easily controllable, is to use
LPCVD with SiH_2Cl_2, NH_3 and N_2O gas inputs, where the N_2O to NH_3 ratio
has been shown[1] to be linear with the O/N ratio in the film, as measured
by Rutherford Back-Scattering (RBS). Similarly, the refractive indices
of the films were dependent[1] on the deposition conditions in an entirely
predictable and reproducible manner, indicating film homogeneity on an
atomic scale.

If silicon oxynitrides are to replace the more conventional
nitride-oxide double layers, then clearly as much as possible must be
learnt about correlations between composition and electrical properties.
There have been many studies of the compositions of oxynitride thin
films, some of the surface as a function of preparation parameters[5-7],
and others as a function of depth into a film[1,2,5-8]. The only attempts
at using surface analytical techniques to study co-ordination in such
films have been those of Streb and Hezel[5], and Hezel and Streb[7], using
Auger Electron Spectroscopy (AES). They followed the changes in the
position of the Si LVV peak and its associated fine structure in the
differential dN(E)/dE spectrum, as a function of oxygen fraction O/(O+N)
in oxynitride films formed by simultaneous implantation of oxygen and
nitrogen ions, and stated that their results confirmed the random
bonding model described later. However, the use of electron-excited AES
in the differential mode, particularly on silicon compounds, in an
attempt to extract chemical information, can pose serious problems. The
latter include (a) the effects of specimen charging under the incident
electron beam, always a problem with silicaceous materials, (b) the
likelihood of damage by the incident beam, changing both the composition

and the chemistry, (c) the use of Auger "peaks" (actually minima) in the differential distribution, when discussing peak shifts, broadening, appearance and disappearance of fine structure, etc., and, with particular reference to the silicon LVV Auger region (d) the presence of a more or less steeply sloping inelastic background under the Auger signals, that makes quantification difficult.

For the above reasons it is preferable to use X-ray photon excitation, as in X-ray Photoelectron Spectroscopy (XPS), where beam damage is avoided, chemical shifts, broadening, fine structure, etc., are much easier to disentangle, and quantification is straightforward. The problem of charging still remains, although not as severely as in AES. To avoid the problem, and also to provide the desired physical insight into the local electronic surroundings of the silicon atoms in the oxynitrides, it is better to measure the difference in energy of certain peaks recorded in the same spectrum, in the form of the Auger Parameter. The latter, conceived originally by Wagner[9], and then subsequently modified by him[10], is defined as

$$\alpha^* = E_K \text{ (Auger)} + E_B \text{ (photoelectron)} \qquad (1)$$

where E_K (Auger) is the kinetic energy of a conveniently situated Auger peak of sufficient intensity, and E_B (photoelectron) is the binding energy of a prominent photoelectron peak. If the ionised core level giving rise to both the Auger transition and the photoelectron ejection is the same, then to a good approximation

$$\Delta\alpha^* = \Delta R^{ea} \qquad (2)$$

where R^{ea} is the extra-atomic relaxation energy associated with the creation of two holes at the ionised atom site. Since R^{ea} depends on the nature of the electronic screening around the atom, it is a good indicator of changes in the chemical environment of that atom.

Although to be rigorous the correct α^* to be measured for silicon should be

$$\alpha^* = E_K \text{ (KLL)} + E_B \text{ (1s)} , \qquad (3)$$

in fact, as shown by Rivière et al.[11], the chemical shifts in silicon 1s are linear with those in silicon 2s and 2p, so that one can substitute $E_B(2s)$ or $E_B(2p)$ in (3) without significantly affecting the relationship (2). The Auger Parameter used here is therefore based on $E_B(2s)$, as used successfully in previous work[12,13]. Since the Parameter varies over more than 2 eV from SiO_2 to Si_3N_4, and can be measured[13] usually to an accuracy of $\pm$ 0.1 eV, any variations with O/N ratio across the oxynitride range should be easily measurable. According to electron diffraction[14], to infra-red[15] and to earlier XPS[16] measurements on oxynitride films grown by Atmospheric Pressure Chemical Vapour Deposition (APCVD), the films are not just mixtures of silicon dioxide and silicon nitride in the appropriate proportions, but are homogeneous on an atomic scale. That is, oxygen replaces nitrogen continuously in the SiN_zO_{4-z} tetrahedra, where z = 0, 1, 2, 3, 4. More recent XPS measurements coupled with X-ray emission spectroscopy[17] have reinforced the random bonding model in which N and O displace one another in a random manner in a disordered network of tetrahedra. However, none of these observations has been able to say what effect the progressive substitution of oxygen for nitrogen has on the chemical surroundings of the silicon, or whether there was a greater influence of oxygen over nitrogen, or vice versa. Such information is inherently available in AES, but its extraction would require very detailed data processing of high quality spectra in the undifferentiated mode; no one has yet attempted that for the oxynitrides. In addition to the need for knowledge about the relative influences on silicon of the oxygen and nitrogen, it is interesting to note that at certain O/N ratios across the range singular properties of the films appear. Thus for O/N ≈ 0.4-0.5, more hydrogen is retained[18] after high temperature annealing than at lower or higher ratios; at O/N ≈ 0.65 a strong reduction in the interface state density is found[19], while for O/N > 0.25 injected charge retention time increases[19] significantly. It is also known[20] that near O/N ≈ 0.3 the resistance of oxynitrides to oxidation is optimum.

In this work X-ray Excited Auger Electron Spectroscopy (XAES) and XPS have been applied together to the study of ionic screening effects in thin, amorphous silicon oxynitride films prepared by the LPCVD method, by measurement of the Auger Parameter during ion bombardment and as a function of O/N ratio. The range of oxynitride compositions studied was from 0.3 to 3.6 with pure Si_3N_4 and SiO_2 films included as standards at each end of the composition range.

2.   EXPERIMENTAL

The oxynitride films were prepared at Philips Research Laboratories according to the LPCVD method described in Reference 1. They were supported on (100) oriented silicon single crystal wafers and varied in thickness from 15 to 100 nm. O/N ratios could be varied in an entirely controllable fashion by changing the N_2O/NH_3 gas phase ratio.

Specimens of area ≈ 1 cm^2 were mounted on standard specimen stubs with copper-backed adhesive tape (3M Corporation), care being taken that none of the tape was exposed to the X-ray photons or argon ion beam. No rise in pressure above the base level of ≈ 5 x10^{-10} mbar was observed on introduction into the spectrometer. XPS and XAES measurements were made in a VG ESCALAB 2 spectrometer, using unmonochromatised Al K_α radiation for the Si 2p and 2s, and the O, N and C 1s photoelectron lines, and Bremsstrahlung radiation from the Al source to excite the Si $KL_{2,3}L_{2,3}$ Auger line. Spectra were recorded at 20 eV pass energy (energy resolution ≈ 1.1 eV), at a source power of 300 W.

For the measurement of surface composition, particularly the O/N atomic ratio, photoelectron peak areas were measured after subtraction of a linear background. Normally a modified[21] Shirley[22] background would have been used, but it so happened that all the photoelectron peaks of interest here were situated on backgrounds so flat that there was no significant difference between the results of subtraction of each type. Relative sensitivity factors for O and N were derived from thin film oxide and nitride standards prepared in the same way as for the oxynitride films. Rather than assume that the standards were stoichiometric SiO_2 and Si_3N_4 their bulk compositions were measured very accurately by Elastic Recoil Detection Analysis (ERDA), from which it was established that the nitrides were indeed Si_3N_4, and the oxides stoichiometric SiO_2. ERDA also revealed that the H concentrations in the standards varied from 0 to 3 at.%.

The oxynitride films were depth profiled with a Leybold-Heraeus IQE 12/38 ion gun, operated normally at 4 keV and $\approx$ 30 $\mu A\ cm^{-2}$. However, as to be described later, because of uncertainties about the effects of ion bombardment on both the elemental composition and the chemistry of oxynitrides, a series of measurements was made on selected oxynitride compositions using also ion energies of 0.5, 1.0 and 2.0 keV, at the same ion doses. The ion beam, of spot size $\approx$ 0.5 mm, was rastered over the surface to give a crater of dimensions 8 x 8 mm, in the centre of which a circular area of approximate diameter 5 mm was used for analysis.

Since measurement and use of an Auger Parameter eliminates referencing problems, no attempt was made to use a flood gun; it is not certain that any significant advantage would have accrued in any case. The particular Auger Parameter used here was

$$\alpha\ =\ E_K\ (Si\ KL_{2,3}L_{2,3})\ -\ E_K\ (Si\ 2s) \qquad (4)$$

which is the same as that already used successfully by Moseley et al.[12], Rivière and Crossley[13], and Crossley and Rivière[23], the range of values of α for several silicon compounds having been measured in Reference 13.

3.   RESULTS

3.1   Effects of Ion Bombardment

Many authors have studied the effects of ion bombardment on films of silicon nitride[24-28], silicon oxide[25,29-31] and silicon oxynitrides[1,6-8,25,31] in order to be able to take account of any such effects in the depth profiling of thin films of these materials. The general finding is that all these materials are affected to a greater or lesser extent by ion bombardment, almost always using argon ions, in that nitrogen or oxygen are preferentially removed, and that some of the silicon near the surface is reduced, leading to the formation of Si-Si bonds at the damage threshold[31], or to elemental silicon well beyond the threshold. However, not all workers have detected such damage after bombardment, and it is essential to note that those that have been able to, have without exception used the Si LVV peak in electron-excited AES as their indicator for silicon reduction. The kinetic energy range of this transition is $\approx$ 77-92 eV, according to chemical state, i.e. right at the minimum of $\approx$ 0.4 nm in the dependence[32] of inelastic mean free path on electron kinetic energy. Thus the variation in line shape of the Si LVV as a function of ion bombardment refers to changes in the silicon chemical state in the outer monolayer only. Other workers[24,29], who used the Si 2p photoelectron peak in XPS as their indicator, found no appearance of a shifted peak due to reduced silicon until the substrate was reached, simply because the Si 2p kinetic energy would have been in the range $\approx$ 1150-1380 eV, depending on the exciting X-ray line used. In this kinetic energy range the inelastic mean free paths would be many times greater, and the signal therefore characteristic of the sub-surface layers rather than of the surface layer itself.

During depth profiling in this work, in which both the Si 2s and 2p photoelectron peaks, and the Si KLL Auger peak, inter alia, were monitored, there was again no indication of any reduction of silicon in the film. This is demonstrated by the sequence of spectra in Figure 1 for the Si KLL. There is no significant contribution to the spectrum from reduced silicon visible until the underlying substrate is reached after 27 min. bombardment.

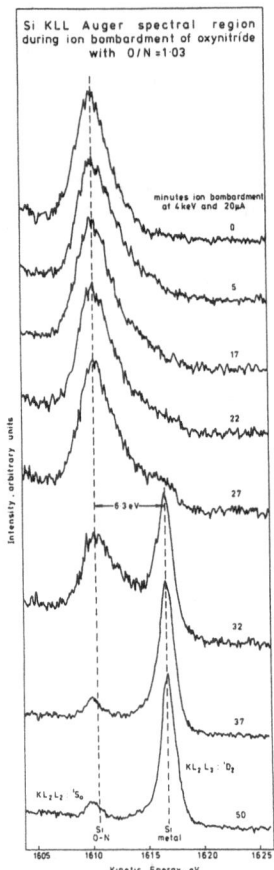

Figure 1   Sequence of spectra in the Si KLL Auger kinetic energy region during ion bombardment of the oxynitride film of composition O/N = 1.03. There is a shift of 6.3 eV between the Si KLL peak positions in the oxynitride and in elemental Si.

The oxynitride film with O/N = 1.03 was of thickness 101.6 nm, and the time taken, $\approx$ 30 min., to penetrate to the substrate with an ion current density of $\approx$ 30 μA cm^{-2} is in line with the sputtering rate measurements of Kuiper et al.[1]. For that composition they found a sputtering rate of $\approx$ 0.68 nm min^{-1} at a current density of $\approx$ 10 μA cm^{-2}

and an ion energy of 2 keV, which would result in removal of ≈ 61 nm for the above duration and current density. The difference of ≈ 40% may be accountable on the basis of the different ion energies used, 4 keV as against 2 keV, and on the different ion incident angles used, but very little seems to be known about the effect of ion energy on sputtering of oxynitrides. Nearly all workers have operated at one chosen ion energy, the only exception being Fransen et al.[27], who used 2 keV and 4 keV.

Because there is so little information about the influence of ion energy on sputtering of oxynitrides, and because ion bombardment must be used for cleaning and depth profiling, the possible influence was studied as an integral part of this work. Four ion energies were chosen, 0.5 keV, 1.0 keV, 2.0 keV and 4.0 keV, and the ion doses at each energy and each step were kept constant. Since such measurements are tedious and time-consuming, particularly at 0.5 keV, two compositions only, O/N = 0.69 and O/N = 1.03, were selected for study. At each step the O/N ratio and the aforementioned Auger Parameter were measured in the ways described earlier.

The results are given in Figures 2 and 3, in which the O/N ratio and the α values, respectively, are plotted against the ion dose on a

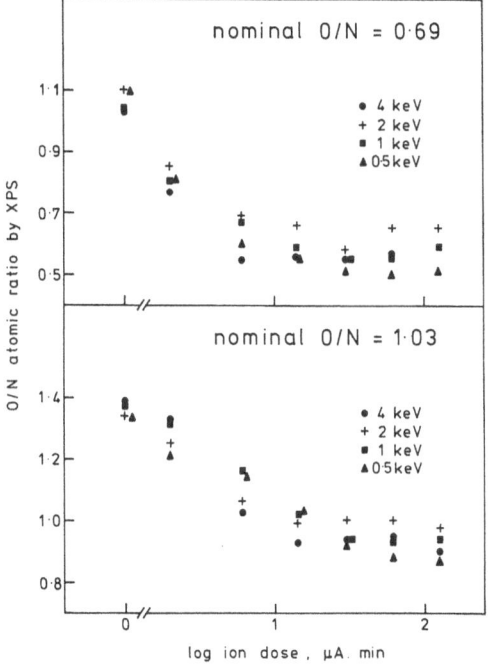

Figure 2   The variation with ion dose of the O/N ratio as measured by XPS, in two oxynitride films of compositions O/N = 0.69 and 1.03 according to RBS/ERDA analyses, for four different ion energies in the range 0.5 keV to 4.0 keV.

logarithmic scale, with the as-received (zero dose) values included on a
broken scale. For both films there was excess oxygen at the surface as
a result of surface oxidation, that was removed quickly by the
bombardment; it should be noted that the slight carbon contamination
also present initially disappeared as well, after the first bombardment.
The O/N ratio in each case then reached a steady-state value after
10-20 µA min. For these two films the steady-state XPS measured values
were some 13% lower than the nominal values, but the general systematic
deviation of the XPS quantification from the RBS/ERDA measured
compositions was 4.5%.

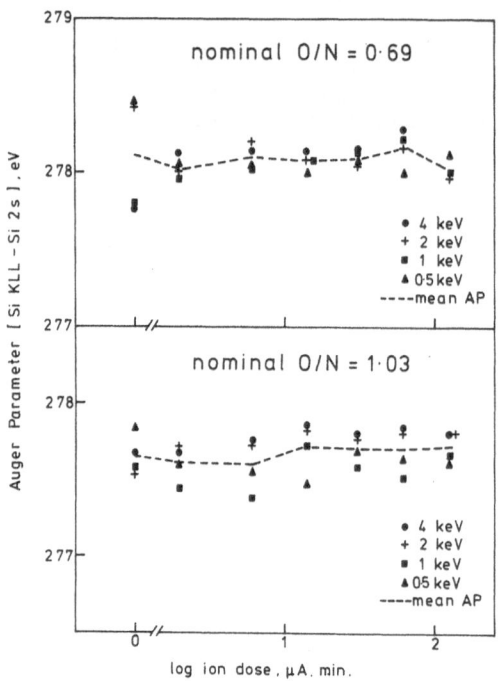

Figure 3 The variation with ion dose of the Auger Parameter
          [Si KLL -Si 2s] as measured by XPS and XAES, in two oxynitride
          films of compositions O/N = 0.69 and 1.03 according to RS/ERDA
          analyses, for four different ion energies in the range 0.5 keV
          to 4.0 keV.

     As far as the effect of different ion energies is concerned, the
data in Figure 2 show that within a maximum experimental error of 10%,
there was no difference from 0.5 keV to 4.0 keV on the shapes of the
profiles with respect to the O/N ratio. There is a suggestion that once
the steady-state condition has been reached, 0.5 keV ions cause
marginally greater depletion of oxygen than ions at higher energies, but
more extensive experiments would be needed to substantiate that.

3.2 Depth Profiling

All the films were depth profiled by alternate ion bombardment and surface analysis, under the standard conditions mentioned earlier. A typical profile of a film of O/N ratio 1.03 and thickness 101.6 nm is shown in Figure 4. The profile of a layered structure, a film consisting of 80 nm of nitride on 30 nm of oxide, is also shown in Figure 5. Although not strictly part of the series described earlier, it contains several interesting features that are relevant here. Small amounts of contaminant carbon were present on the surfaces before bombardment, but have been ignored in deriving the compositions entered in the Figures; the carbon disappeared after the first bombardment in any case. In addition to the atomic percentages of Si, O and N, the values of α measured at each stage of bombardment are shown, with the values[13] for SiO_2, Si_3N_4, and Si on the same scale for comparison.

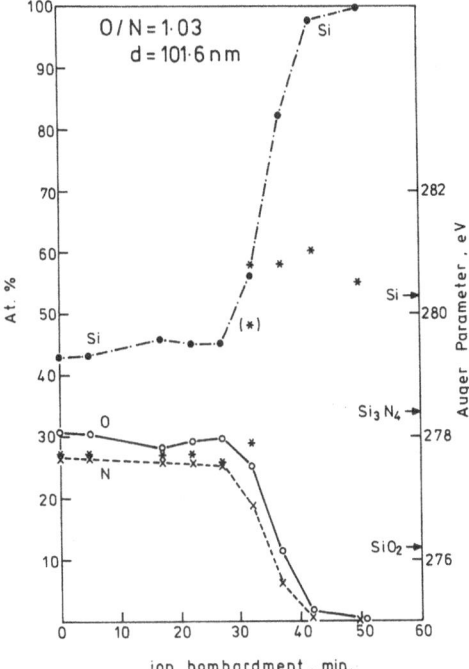

Figure 4   Depth profile with 4 keV argon ions through an oxynitride film of thickness 101.6 nm and composition O/N = 1.03 given by RBS/ERDA analysis. Also entered are the measured Auger Parameter values (stars) through the film, with the values for bulk SiO_2, Si_3N_4, and Si, for comparison. The AP value in brackets at the film/substrate interface has been derived from curve resolutions of the Si 2p and KLL spectra.

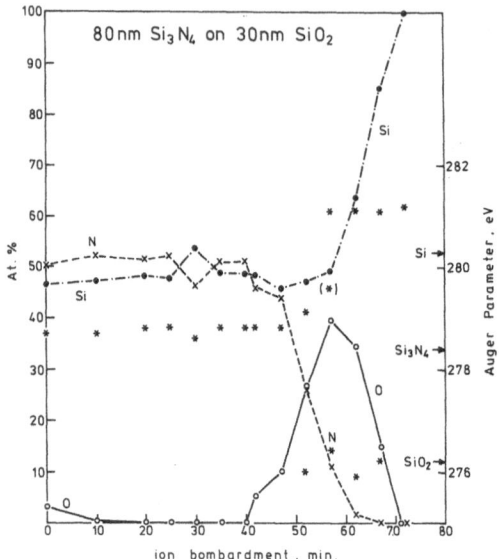

<u>Figure 5</u> Depth profile with 4 keV argon ions through a layered film
structure of 80 nm Si_3N_4 on 30 nm SiO_2 on a Si substrate.
Also entered are the measured Auger Parameter values (stars)
through the film, with values for bulk SiO_2, Si_3N_4, and Si,
for comparison. The AP value in brackets at the nitride/oxide
interface was derived by inspection of the Si 2s and KLL
spectra in Figure 7.

There was a tendency for the surface to contain a slight excess of
oxygen. The O 1s peak was always a single peak, with no apparent OH
contribution, so evidently the excess oxygen was due to additional
oxidation of the oxynitride, and not to any pick-up of water.

The oxynitride/silicon interfaces in the films appear broad, but
much of the broadening will have been due to the progressive degradation
of depth resolution with depth into the film as a statistical
consequence[33] of the ion bombardment process itself. A maximum
interface width of 2.5 nm can be derived statistically; the likelihood
is in fact that it is abrupt on an atomic scale. Similarly the
nitride/oxide and oxide/silicon interfaces in the layered film
(Figure 5) are likely to be much narrower than they appear,
although it is difficult from the data in that Figure to derive values
for the maximum widths.

During profiling the spectral behaviour was in all cases that,
first of the appearance of shoulders in the Si 2s, 2p and KLL spectra in
the elemental Si position, then of the development of the shoulders into
sharp peaks with continued bombardment while the oxynitride peak
decreased, and finally of the disappearance of the oxynitride peak,
<u>without</u> any energy shift in the latter. However, in favourable
situations, i.e. where the surface exposed at a particular step in

profiling happened to produce spectra in which the oxynitride and
elemental contributions were comparable, curve resolution of the spectra
allowed small additional interface structures to be detected. One such
situation can be seen in Figure 1 at the 32 min. step. The curve
resolution carried out on the Si KLL Auger spectra at that point in the
profiling is shown in Figure 6. Following such curve resolution, small
residual intermediate peaks were found in both 2p and KLL spectra. If
the reasonable assumption is made that they originated from the same
state of co-ordination of silicon then an Auger Parameter can be derived
(knowing that the 2p-2s separation is exactly 51.0 eV[13]) which is itself
intermediate between the two derived from the major peaks. This third,
less accurate, α value has been entered in Figure 4 in brackets at the
32 min. bombardment step.

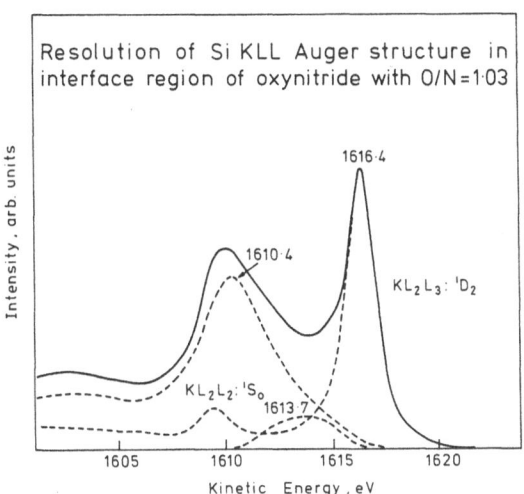

Figure 6   Curve resolution of the Si KLL Auger spectrum in the interface
region (i.e. after 32 min. ion bombardment) of the oxynitride
film with composition O/N = 1.03 given by RBS/ERDA analysis.
A small residual peak at $\approx$ 1613.7 eV was left after the
various subtraction procedures.

Even more complex spectra, involving possibly more than one
intermediate state, were recorded from the nitride/oxide interface in
the layered nitride-on-oxide film. The Si 2s and KLL spectra from this
region, at the 57 min. ion bombardment stage, are shown in Figure 7.
Although no curve resolution was attempted on these spectra due to the
inferior signal-to-noise quality of the KLL spectrum, at least three
contributions to the latter seem to be present. Associating the 2s
contribution at $\approx$ 1333.3 eV with the KLL contribution at $\approx$ 1612.9 leads
to the third α value entered in brackets in Figure 5.

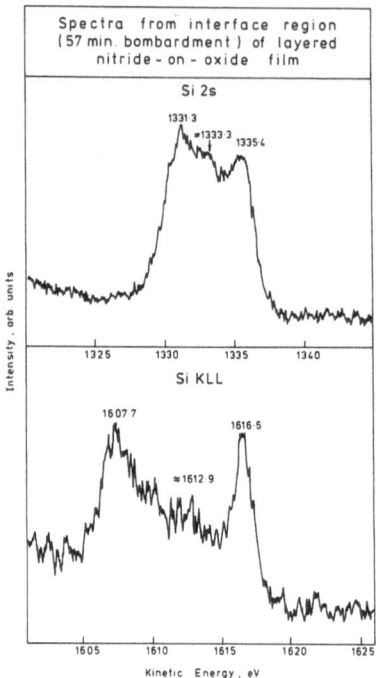

<u>Figure 7</u>   The Si 2s and KLL Auger spectral regions recorded at the
nitride/oxide interface in the layered nitride-on-oxide-on-
silicon film. No data processing has been performed, but
additional structure in both spectra can be discerned between
the oxide and nitride energetic positions.

3.3 Line Widths and Shapes

The widths of the Si 2s, 2p and KLL peaks, based on measurement of
the full-width-at-half-maximum (FWHM) following background subtraction,
were measured for each oxynitride film and for the oxide and nitride
film standards, at several points during profiling once the O/N ratio
had become constant. That is, referring to Figure 2, after ion doses of
about 30 µA.min. The FWHM values so obtained are shown in Figure 8 for
the Si KLL peaks as a function of oxygen fraction (O/O+N); the
variations for the Si 2s and 2p were similar and are not shown. It can
be seen that the peak width increased from that in the nitride as oxygen
was introduced, went through a broad maximum around an oxygen fraction
of ≈ 0.5, and decreased again to that in the oxide.

The nature of the broadening is shown in Figure 9, in which the KLL
peak at the O/N ratio corresponding to maximum peak width is compared
with the same spectral feature from the nitride film. For the
comparison purposes, the peak has been shifted as required to achieve
alignment along its centroid line, that is, at the same kinetic energy,
and normalised to the same intensity. It is clear that the broadening
is almost the same on each side with no obvious additional structure.

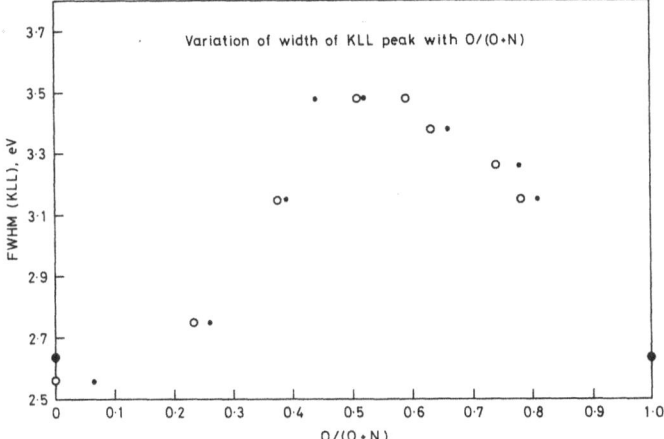

Figure 8   Variation of the width (FWHM) of the Si KLL Auger peak as a
function of the oxygen fraction O/(O+N) in the oxynitrides.
Full circles – RBS/ERDA quantification; open circles – XPS
quantification.

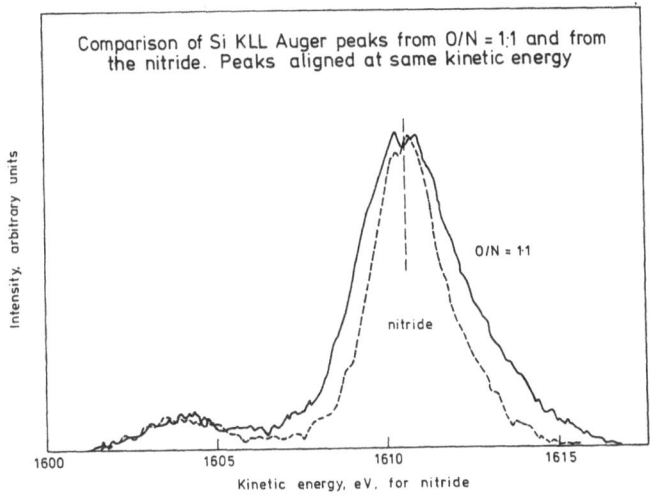

Figure 9   Comparison of the Si KLL Auger peaks from the oxynitride
composition, O/N = 1.1 (i.e. O/(O+N) = 0.52), having the
greatest KLL FWHM, and from the nitride. For comparison
purposes the oxynitride peak has been shifted in energy so
that the peak centroids are aligned.

Apart from the broadening, there was no change in basic peak shape on traversing the range of compositions from nitride to oxide.

3.4 Auger Parameter Variation with O/(O+N)

As with the FWHM, values of the silicon Auger Parameter $[E_K(KLL)-E_K(2s)]$ were taken to be characteristic of a particular composition beyond the point in ion profiling at which the XPS-measured O/N ratio became sensibly constant. These values are plotted as a function of the oxygen fraction, O/(O+N), derived from XPS quantification, in Figure 10. Within the scatter of experimental error (almost entirely in the quantification and not in the α measurements) it can be seen that the α values decreased monotonically but apparently not linearly from Si_3N_4 to SiO_2. There are no obvious discontinuities, although to be absolutely certain additional compositions in the range $0.8 \leq O/(O+N) \leq 1.0$ should be studied. The straight line drawn in the Figure shows the dependence of α to be expected if the oxynitride films

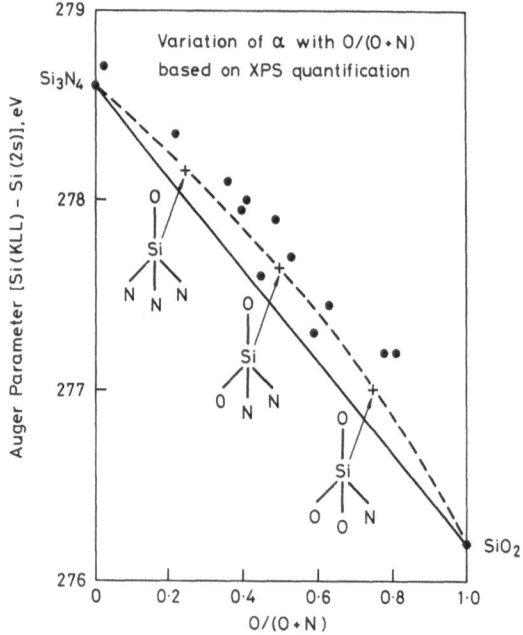

Figure 10 Variation of the Auger Parameter [Si(KLL) − Si(2s)] with the oxygen fraction O/(O+N) based on XPS quantification, in the oxynitrides across the range $0 \leq O/(O+N) \leq 1$ (full circles). The straight line is the dependence to be expected if the oxynitrides were merely proportionate mixtures of separate Si_3N_4 and SiO_2 phases. The curved dashed line shows the dependence to be expected from expression (6), assuming the tetrahedral compositions indicated, and the normal 3 and 2 valencies for N and O, respectively.

consisted merely of variable mixtures of nitride and oxide, and it is clear that the observed dependence is considerably displaced from the linear.

It is hardly surprising that the relationship between α and composition is non-linear. The generally accepted model is that of progressive substitution of nitrogen by oxygen in the SiO_xN_y tetrahedron, in which, for nearest-neighbour interaction only, there would be three possible intermediate configurations. The expected changes in co-ordination of the silicon in this model, i.e. the changes in α, can then be calculated for each configuration, assuming nearest-neighbour dependence only, and taking the normal valencies of 2 for oxygen and 3 for nitrogen. With α for the nitride film measured as 278.6 eV, and with a difference of 2.4 eV between nitride and oxide, one then has the formula

$$\alpha = 278.6 - 2.4 \left\{ 2x/[2x + 3(1-x)] \right\} \text{ eV} \qquad (5)$$

that is, $$\alpha = 278.6 - 4.8x/(3-x) \text{ eV} \qquad (6)$$

where $$x = O/(O+N).$$

When α is calculated from (6) for each of the three intermediate configurations, the values are seen to fall on a bow-shaped curve, drawn dashed in Figure 10, with the three configurations indicated. The bow shape is a consequence of the different valencies of oxygen and nitrogen.

The interesting observation is that the experimentally derived α values generally fall outside the theoretical curve, i.e. in the direction of higher α than would be expected on the above simple model. Now the assumption was made implicitly in the model that the respective contributions of oxygen and nitrogen to the extra-atomic relaxation energy of silicon were the same in the oxynitrides as in the nitride and the oxide. If it is assumed that oxygen and nitrogen may make different contributions, and if this difference is expressed as a ratio n, where n = 1 for equal contributions, and n > 1 for a greater contribution from nitrogen vis-à-vis oxygen, then equation (6) may be re-written

$$\alpha = 278.6 - 2.4 \left\{ 2x/[2x + 3n(1-x)] \right\} \qquad (7)$$

Application of a least-squares best fit to the data in Figure 10 produces the result that n = 1.20 ± 0.1. In other words, considering nearest-neighbours only, the effect of nitrogen on the extra-atomic relaxation energy of silicon in the oxynitrides is apparently some 20% greater than that of oxygen.

4.   DISCUSSION

4.1   Ion Bombardment Effects and Ion Profiles

The spectral sequence of Figure 1, taken in conjunction with the results of Figure 3 for the variation of Auger Parameter during bombardment with a range of ion energies, have important practical consequences. The first conclusion to be drawn is that there was no free silicon in the oxynitride films, since results similar to those for

O/N = 1.03 in Figure 1 were obtained for all compositions. The second conclusion must be that the use of the Si LVV peak (particularly electron-excited, for the reasons given earlier) both as a measure of silicon concentration and as an indicator of the approach to the substrate during depth profiling, should be avoided. This is in complete agreement with the statement by Jung and Titel[28] that "The Si LVV line is of very limited use for quantitative analysis". As remarked earlier, many workers have found ion bombardment effects in silicon oxynitrides, especially that of the reduction of silicon, but always when using the Si LVV peak. Electron irradiation effects are also readily observable[7] at the oxygen-rich end of the O/N range. Here, however, and in other work[24,29] using the Si 2p photoelectron peaks at much higher kinetic energies than the LVV transition, no artefacts introduced by ion bombardment were observed. This is in line with the suggestion of Thomas and Hofmann[29] that ion beam damage is confined to a thickness corresponding to the Si-O bond length, i.e. ≈ 0.16 nm, or about one atom layer, to which the LVV transition at 77-92 eV would be very sensitive, but which would hardly be noticed at the kinetic energies of the Al-excited 2p, 2s and KLL peaks. The conclusions of Paparazzo[30] are implicitly the same. The synergistic effects of combined electron and ion irradiation, as happens during AES multiplexing, are likely to be even more catastrophic.

The above conclusion is supported by the dependence of Auger Parameter on ion dose shown in Figure 3. Although the data in Figure 2 show that the surfaces of the two films were oxidised, the values of α found before bombardment started were not characteristic of an oxidised surface (cf. α ≈ 276 eV for SiO_2) but of the oxynitride itself, since they were the same as found in the bulk of the film. This invariance, coupled with the virtual lack of dependence on ion energy, confirms that measurement of the Auger Parameter used here is unaffected by ion beam damage to the first atom layer, at least for ions with energies between 0.5 and 4.0 keV. The latter result is important in the present context since it means that this Auger Parameter can be used to study any variations in co-ordination through an oxynitride film, and partly at least across the interface with the substrate, without having to worry about the effects of ion bombardment. It must therefore be recommended that depth profiling of silicon-containing materials be performed using spectral features at high kinetic energy and avoiding electron irradiation, certainly simultaneous irradiation, if possible.

The depth profiles in Figures 4 and 5 show several interesting features. Firstly, the observation that the Auger Parameter remains unchanged at the value characteristic of the particular O/N ratio, right up to the Si-O-N/Si interface or, in the case of the layered structure of Figure 5, right up to the Si_3N_4/SiO_2 interface, confirms the indication that the interfaces are probably sharp. If any intermixing had occurred at the interfaces then the α values found through the film would have changed measurably as the interface was approached. No such shift was seen, although, as shown in Figures 4 and 5, based on curve resolution operations similar to those in Figure 6, additional co-ordination states could be found at those profile points in interfaces where the intensities of the Si photoelectron and Auger peaks allowed such curve resolution to be performed sensibly. Such intermediate states have been observed in the interface regions of other thin films on Si, especially of SiO_2 on Si and it is likely that the same explanation for their origin in SiO_2/Si interfaces will apply here. Although the interface states found in this work appeared following depth profiling by ion bombardment, whereas in the SiO_2/Si interface they were observed by angular resolved measurements, it should be emphasised again that the Auger Parameter used here is based on spectral features at high kinetic energy which do not show ion damage effects in the oxynitride itself.

A second interesting observation is that immediately after the interface in each film the Auger Parameter is significantly higher than the value[13] for bulk elemental Si, but shows a tendency in a few instances to drop towards the latter on continued bombardment. The higher value is similar to that found (ref. 13 and other refs. therein) for silicides, although there is no possibility of silicide formation in the systems studied here. The precise reason for this observation remains uncertain at the moment.

Thirdly, for nearly all the films the elemental and α profiles were reasonably flat, indicating that both constant composition and constant co-ordination had been achieved throughout the film by the deposition process. The only exception was that of a film with O/N = 1.42 as measured by RBS/ERDA, in which the elemental profiles were not flat but changed progressively with depth, so that the O/N ratio measured by XPS varied continuously from 0.6 to 1.7–1.9 near the interface. The average O/N ratio was close to the nominal. Several samples with the same O/N content were examined, with the same result, and no conclusion could be drawn apart from the obvious one that the deposition conditions might themselves have been changing during film formation.

4.2 Peak Widths

Peak broadening can arise from several different physical sources, but when, as here, the only parameter that is changing is an elemental ratio, the most likely source is that of contributions from additional chemical states. These states might be thought to correspond formally to the five possible nearest–neighbour configurations in the silicon tetrahedron as nitrogen is substituted by oxygen, but with an overall shift of $\approx$ 4 eV in the Si KLL position from nitride to oxide, the individual contribution from each configuration should have been easily resolvable. Instead the positions of both Auger and photoelectron peaks shifted in a continuous fashion with O/N ratio, without any additional structure, which implies that the five formal configurations did not make specific individual contributions, and with the further implication that the nearest–neighbour model is therefore inadequate. Next–nearest–neighbour or even more remote shells might need to be considered.

The observed peak broadening in KLL, 2s and 2p peaks must therefore have arisen from other chemical effects, and since the broadening was both structureless, and uniformly distributed around the peaks (Figure 9), the additional chemical states must have contributed a quasi–continuous distribution of energies to each side of the peak centre. Such a distribution of closely spaced energies is characteristic of a variable defect structure. Any departure from stoichiometry would have been too small to be measured, and the Auger Parameters were derived in any case from the energies at peak maxima, so the broadening is the only indicator in this type of measurement of the possible existence of a defective structure in the oxynitride films. "Defective" in this context is used in a general sense; it includes not only genuine lattice defects but also a range of micro-effects such as progressive changes with O/N of the distribution of Si-O-N bond angles and of Si-O, Si-N, bond distances.

Although the above interpretation cannot be confirmed here, and the precise nature of the defects, if they exist, is unknown, it is interesting to note that it is in just the composition region, O/(O+N) $\approx$ 0.5, in which the maximum broadening is observed, that the oxynitride films show some singular properties, as pointed out in the Introduction. Those that might well be dependent on the nature and extent of any defect structure are the interface state density, which shows[19] a reduction around O/(O+N) $\approx$ 0.4, and the injected charge retention time, which increases[19] significantly beyond O/(O+N) $\approx$ 0.2. The ability to retain hydrogen in the film, which is maximum[18] around O/(O+N) $\approx$ 0.3, might also be a function of the precise nature of any defects.

4.3 Auger Parameter Variation

Within an experimental error of 15%, the data of Figure 10 show that there is an apparently smooth variation of Auger Parameter with composition across the range from Si_3N_4 to SiO_2. There are no obvious indications of any discontinuous jumps from an α value characteristic of one particular structure to that of the next. This observation is entirely in accord with the generally accepted model of a progressive substitution of nitrogen by oxygen in the SiO_xN_y tetrahedra, rather than that of a mixture of separate SiO_4- and SiN_4-tetrahedra. It is worth emphasising at this point, however, that although the above conclusion is only a confirmation of what is already accepted on the basis of previous measurements[14-16], had there been no such measurements then the approach adopted here would have been able to state categorically that the progressive substitution model was the correct one.

5. CONCLUSIONS

Measurement of the Auger Parameter $[E_K(Si\ KL_{2,3}L_{2,3}-E_K(Si\ 2s)]$ has been used to study a variety of effects in silicon oxynitride films prepared by LPCVD. Compositions used varied from O/N = 0.3 to O/N = 3.6 as well as the end members Si_3N_4 and SiO_2, and the films had thicknesses between 20 and 150 nm. The following detailed conclusions have been reached:

1. During bombardment of silicon oxynitrides with argon ions of energies in the range 0.5 keV to 4.0 keV there was no dependence of either composition or co-ordination (via the Auger Parameter) on ion energy.

2. Ion bombardment damage to oxynitrides in the above energy range is restricted to the first atom layer. By using the high kinetic energy Si 2s, Si 2p, and Si KLL Auger emissions, both compositional analysis and measurement of the co-ordination state via the Auger Parameter can be performed free from ion beam induced artefacts, even during depth profiling.

3. Although there is plenty of evidence of electron beam damage during analysis of oxynitrides, there was no evidence of any degradation by the soft X-ray photon irradiation in XPS.

4. The lack of both ion-induced and electron-induced artefacts in the approach used here to study the oxynitrides has allowed the conclusions to be reached that there was no elemental Si present in the films, and that the oxynitride/silicon-substrate interfaces were probably sharp.

5. Quantification of film composition by XPS was straightforward. The resultant O/N ratios were on average $\approx$ 4.5% lower than those measured by RBS and ERDA.

6. Curve resolution of the Si 2p and Si KLL spectra in the oxynitride/ substrate interface region of some films revealed additional chemical states of Si, although their exact nature could not be determined.

7. Although the film surfaces were oxidised, the layer containing additional oxygen must have been very thin, since the Auger Parameter measured before ion bombardment commenced was characteristic of the particular O/N ratio in the bulk of the film. Similarly the characteristic Auger Parameter remained unchanged right up to the oxynitride/substrate interface, emphasising again the utility of the approach.

8. Immediately after the interface the Auger Parameter was found to have a higher value than that for elemental Si, but on further bombardment into the substrate showed a tendency to drop to the latter value.

9. The widths of the Si 2s, Si 2p, and Si KLL peaks were found to go through maxima as a function of O/(O+N) between Si_3N_4 and SiO_2, at O/(O+N) $\approx$ 0.5. The broadening is suggested as arising from an increased number of chemical states very closely spaced in energy, and therefore not resolvable individually. They may arise from an increase in the defective nature of the films at those compositions.

10. The Auger Parameter appeared to vary monotonically with O/(O+N), confirming the accepted model of random progressive substitution of N by O, without the formation of separate phases. However, the monotonic dependence was found to deviate significantly from the theoretical dependence in the direction of higher Auger Parameter, suggesting a greater influence of N on the extra-atomic relaxation energy compared to that of O.

ACKNOWLEDGEMENTS

    The research described in this paper was funded jointly by the CEC under ESPRIT Project 369 and by the UKAEA Underlying Research Programme. The authors are grateful to the other members of the Project 369 Committee for many helpful discussions, and in particular to Drs. Ton Kuiper and Frans Habraken for critical readings of the manuscript. Dr. Kuiper is also thanked for arranging provision of the LPCVD oxynitride samples.

REFERENCES

1. A.E.T. Kuiper, S.W. Koo, F.H.P.M. Habraken and Y. Tamminga, J. Vac. Sci. Tech., B1, 62 (1983).

2. F.H.P.M. Habraken, A.E.T. Kuiper, Y. Tamminga and J.B. Theeten, J. Appl. Phys., 53, 6996 (1982).

3. R. Hezel, T. Meisel and W. Streb, Proceedings of the 13th European Solid State Device Research Conference, Canterbury, UK, 1983 (Institute of Physics, London and Bristol), E.H. Rhoderick, ed; Europhysics Conference Abstracts, F7, 1983.

4. R. Hezel, T. Meisel and W. Streb, J. Appl. Phys., 56, 1756 (1984).

5. W. Streb and R. Hezel, J. Vac. Sci. Tech., B2, 626 (1984).

6. Y. Yoriume, Thin Solid Films, 115, 135 (1984).

7. R. Hezel and W. Streb, Thin Solid Films, 124, 35 (1985).

8. A. van Oostrom, L. Augustus, F.H.P.M. Habraken and A.E.T. Kuiper, J. Vac. Sci. Tech., 20, 953 (1982).

9. C.D. Wagner, Anal. Chem., 47, 1201 (1975).

10. C.D. Wagner, L.H. Gale and R.H. Raymond, Anal. Chem., 51, 466 (1979).

11. J.C. Rivière, J.A.A. Crossley and G. Moretti, AERE R-12689 (1987).

12. P.T. Moseley, G. Tappin, J.A.A. Crossley and J.C. Rivière, Corrosion Science, 23, 901 (1983).

13. J.C. Rivière and J.A.A. Crossley, Surface Interface Analysis, 8, 173 (1986).

14. N.C. Tombs, F.A. Sewell and J.J. Comer, J. Electrochem. Soc., 116, 862 (1969).

15. M.J. Rand and J.F. Roberts, J. Electrochem. Soc., 120, 446 (1973).

16. S.I. Raider, R. Flitsch, J.A. Aboaf and W.A. Pliskin, J. Electrochem. Soc., 123, 560 (1976).

17. I.A. Brytov, V.A. Gritsenko and Yu.N. Romashchenko, Sov. Phys. JETP, 62, 321 (1985).

18. F.H.P.M. Habraken, R.H.G. Tijhaar, W.F. van der Weg, A.E.T. Kuiper and M.F.C. Willemsen, J. Appl. Phys., 59, 447 (1986).

19. J. Remmerie, H.E. Maes, M. Heyns, R. de Keersmaecker, F.H.P.M. Habraken, J.B. Oude Elferink, W.F. van der Weg and A.E.T. Kuiper, Electrochem Soc. Extended Abstracts, 86, 556 (1986).

20. A.E.T. Kuiper, M.F.C. Willemsen, J.M.G. Bax and F.H.P.M. Habraken, to be published in Appl. Surf. Sci., (1988).

21. H.E. Bishop, Surface Interface Analysis, 3, 272 (1981).

22. D.A. Shirley, Phys. Rev. B, 5, 4709 (1972).

23. J.A.A. Crossley and J.C. Rivière, Corrosion Science, 29, 45 (1989).

24. T.N. Wittberg, J.R. Hoenigman, W.E. Moddeman, C.R. Cothern and M.R. Gulett, J. Vac. Sci. Tech., 15, 348 (1978).

25. R. Hezel and N. Lieske, J. Appl. Phys., 51, 2566 (1980).

26. R. Hezel and N. Lieske, J. Appl. Phys., 53, 1671 (1982).

27. F. Fransen, R. Vanden Berghe, R. Vlaeminck, M. Hinoul, J. Remmerie, and H.E. Maes, Surface Interface Analysis, 7, 79 (1985).

28. T.Jung and W. Titel, phys. stat. sol. (a), 98, 63 (1986).

29. J.H. Thomas III and S. Hofmann, J. Vac. Sci. Tech., A3 , 1921 (1985).

30. E. Paparazzo, J. Phys. D: Appl. Phys., 20, 1091 (1987).

31. S.S. Chao, J.E. Tyler, D.V. Tsu, G. Lucovsky and M.J. Mantini, J. Vac. Sci. Tech., A5, 1283 (1987).

32. M.P. Seah and W.A. Dench, Surface Interface Analysis, 1, 2 (1979).

33. M.P. Seah and C.P. Hunt, Surface Interface Analysis, 5, 33 (1983).

Chapter 3

OXIDATION OF LOW-PRESSURE-CHEMICAL-VAPOUR DEPOSITED SILICON OXYNITRIDE FILMS

A.E.T. Kuiper, M.F.C. Willemsen and J.M.L. Mulder

Philips Research Laboratories,

5600 JA Eindhoven, The Netherlands

and

J.B. Oude Elferink, R. Erens, F.H.P.M. Habraken and

W.F. van der Weg

Dept. of Atomic and Interface Physics, Utrecht State University

P.O. Box 80 000, 3508 TA Utrecht, The Netherlands

ABSTRACT

In view of their application as oxidation mask in a modified LOCOS (Local Oxidation of Silicon) technology the oxidation of Low Pressure Chemical Vapour Deposited (LPCVD) silicon oxynitride was investigated. The layer compositions studied ranged from that of nitride to oxynitride with an atomic ratio of O/N=1. Oxidations were performed at 850-1000°C in ambients having different H_2O/O_2 flow ratios. Rutherford Backscattering Spectrometry and Elastic Recoil Detection were employed to quantify the increase in oxygen content of the films upon oxidation. Hydrogen profiles in oxidized samples were measured using Nuclear Reaction Analysis and Elastic Recoil Detection.

We observed that the oxidation rate for either oxynitride composition is nearly two orders of magnitude smaller than that of silicon. Linear oxidation kinetics were observed for the conditions applied in this study, which is indicative of a reaction controlled process. The activation energy of this process was determined to be 48 kcal/mole for Si_3N_4 and decreased for oxynitrides with O/N ratios larger than 0.4.

The hydrogen profiles measured in oxidized samples showed a peak at the oxide/(oxy)nitride interface. The size of this peak varies with oxidation

[1] Reprinted from the Journal of Vacuum Science and Technology (june 1989) with permission.© American Institute of Physics 1989

conditions ·and film composition. A reaction mechanism for the oxidation of (oxy)nitride is proposed that relates the observed hydrogen accumulation to Si-NH and Si-OH groups at the interface.

The measured acceleration of the oxidation rate induced by adding small amounts of HCl to the furnace is interpreted in terms of enhanced transport of NH_x species to the surface. This effect is greatest for the nitride and diminishes with increasing oxygen content in oxynitride material.

I. INTRODUCTION

The continuing demand for smaller dimensions in advanced IC designs requires a compact isolation scheme. A substantial reduction of the lateral encroachment of the field oxide (bird's beak) in the conventional LOCOS (Local Oxidation of Silicon) technique [1] is achievable if an oxynitride layer is used instead of the combination of silicon nitride and stress relief oxide [2]. This modified LOCOS technology is very attractive since it does not add any process step compared to the conventional scheme. For optimum performance deeper insight into the oxidation behaviour of oxynitride films may be useful.

In literature a large amount of work has been published dealing with the oxidation of Si. Although there is general agreement that, apart from the initial stage, the rate of this process is determined by diffusion of oxidizing species to the reaction interface, opinions diverge as to the specific mechanism [3]. Especially on the role of water in accelerating the process of oxidation different theories have been presented, but convincing proof for any of them seems hard to provide [4].

Far less has been published concerning the oxidation of silicon nitride. To our knowledge, no kinetic studies are available on Low-Pressure Chemical Vapour Deposited (LPCVD) Si_3N_4, whereas only a few quantitative studies have been documented for Atmospheric-Pressure CVD (APCVD) Si_3N_4 [5-7]. These studies show that the oxidation of nitride proceeds more slowly in dry oxygen than in a wet ambient [5-6]. The kinetic data presented are indicative of diffusion-controlled process, but simple Deal-Grove behaviour is not observed. Enomoto et al. [6] have proposed a model in which NH_3 is assumed to pile up at the oxide/nitride interface during oxidation and to subsequently suppress the oxidation rate. A similar concept has been proposed by Chramova et al. [7]: a counter-diffusion flux of N_2 or NO through the grown oxide layer is thought to be responsible for the lower oxidation rates found for nitride as compared with silicon. Moreover, they have found that the oxidation rate decreases with decreasing hydrogen content of the nitride film. They have elegantly proved that the limiting process has to be related to the grown oxide layer rather than to hydrogen transport from the bulk of the nitride to the SiO_2/Si_3N_4 interface.

Some information about the oxidation of silicon oxynitride is available due to its application as a ceramic. For instance, Si_2ON_2 is reported to be one of the most stable oxynitrides in oxidizing atmospheres at temperatures of 1100-1400°C, and its oxidation resistance is even better than that of pure Si_3N_4 [18-11]. Kinetic data of the oxidation of oxynitride as a refractory material have been published, but a direct comparison with LPCVD material cannot be made. The hot-pressed ceramics usually have pores or boundaries between crystalline grains, which may play a dominant role in the diffusion of oxidizing species to the (oxy)nitride interface [11,12]. In any case the oxidation in these class of materials is best described by a diffusion-controlled process [10-12].

Since extensive kinetic data for LPCVD oxynitrides were evidently lacking, we investigated the rate of oxidation for different SiO_xN_y film compositions in the range between O/N = 0 and O/N = 1, hence including pure Si_3N_4. Parameters varied in the oxidation process were temperature (850-1000°C), time (0.5-16h) and ambient (dry-wet, with and without HCl).

We have established earlier that LPCVD nitride films oxidize layer by layer starting from the exposed surface, the reaction interface remaining essentially sharp like in silicon oxidation [13]. As a consequence, oxidation rates may be determined from grown oxide thicknesses. In this study we employed the nuclear techniques known as Rutherford Backscattering Spectrometry (RBS) and Elastic Recoil Detection (ERD) to quantify the increase in oxygen content of the layers due to oxidation. A possible role of hydrogen in the oxidation kinetics was explored, applying Nuclear Reaction Analysis (NRA) and ERD to measure hydrogen profiles in oxidized samples.

This paper presents quantitative data on the oxidation of LPCVD silicon nitride and oxynitride films. On the basis of these data a reaction mechanism is proposed which describes qualitatively the oxidation behaviour of silicon (oxy)nitride.

II. EXPERIMENTAL

II-1 Sample preparation

For all samples prepared in this study the same substrate material was used: B-doped (17-23 Ohm.cm) Si, FZ material. The (100) wafers had a diameter of 100 mm and the standard backside damage.

On these substrates oxynitride films were grown according to the LPCVD process described before [14]. The deposition temperature for this process was established at 820 °C. Different batches of oxynitride samples were prepared, with film compositions corresponding to an oxygen-to-nitrogen atomic ratio (O/N) ranging from 0 to 1.0. Film thicknesses were grown between 40 nm and 150 nm. Half of the samples from each batch were given an additional anneal in N_2 at 1000°C for 1 h.

Oxidation runs were performed at 850, 950 and 1000 °C using a standard furnace in a clean room facility. The oxidation ambient was varied such that experiments under 6 different H_2O/O_2 flow ratios in the furnace were performed: H_2O/O_2 = 0/100, 25/75, 50/50, 67/33, 75/25 and 90/10. The total flow into the furnace was maintained at 7500 sccm. The required H_2O/O_2 ratio was obtained by mixing the proper amounts of H_2 and O_2. To give an example: an H_2O/O_2 ratio of 50/50 was achieved by mixing 3750 sccm H_2 with 5625 sccm O_2. Since at the oxidation temperature all H_2 will be converted into H_2O, 1875 sccm of O_2 will be consumed and the flow in the furnace will be a mixture of 3750 sccm H_2O and 3750 sccm O_2. In the following, oxidation in a gas phase composition of H_2O/O_2= 90/10 will be referred to as wet oxidation.

In a number of oxidation runs HCl was added to the ambient by bubbling N_2 through a container of trichloroethane held at 30°C. The HCl concentration in the ambient could be varied between effectively 0.8 and 6 vol % by adjusting the N_2 flow.

Oxidation runs were carried out for 0.5, 1, 2, 4, 9 and 16h. Upon completion of the actual oxidation the furnace was ramped down to 900°C in 25 min., subsequently to 850°C in 17 min., and the boat carrying the wafers was then unloaded from the furnace in 15 min., all under N_2 purge.

In all oxidation experiments a Si wafer and a wafer with a 100 nm thick thermal oxide layer were included for comparison. The oxide thickness, grown on the reference Si wafer, was measured to check the correct performance of all oxidation experiments.

The total number of wafers used for this study exceeded 400.

II-2. Oxygen analysis

The extent of oxidation was derived from RBS and ERD analysis of the samples. Both analytical techniques are quantitative and measure the absolute amount of N and O atoms per cm^2, not needing any further calibration.

RBS

As an example, RBS spectra of an oxynitride film (O/N=0) on Si, before and after oxidation, are given in fig. 1. The spectra were invariably

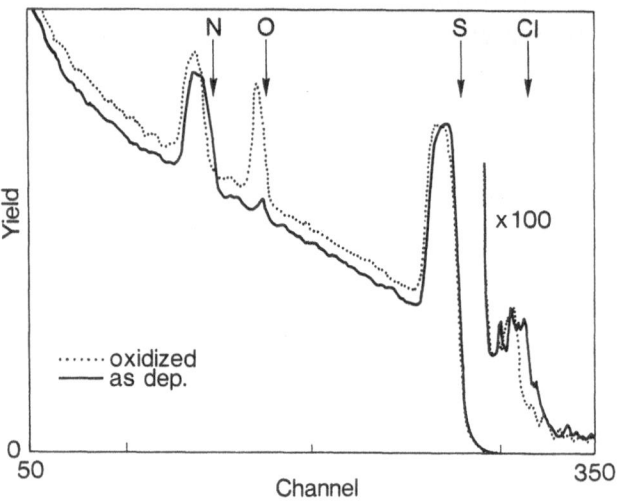

Fig. 1. RBS spectra of 40 nm Si_3N_4, before and after 9 h oxidation in H_2O/O_2 =50/50 at $1000^{o}C$. The arrows identify the surface energy positions of the elements indicated.

recorded under channeling condition in order to reduce the background due to the Si substrate, which increases the accuracy of the determination of the N and O peak areas. Comparison of the spectra in fig. 1 shows that oxidation of silicon nitride proceeds from the outer surface: the oxide forms at the surface and the nitrogen signal is found below the surface in the oxidized sample (see arrows). This layer-by-layer oxidation is typical for (oxy)nitride films, as evidenced by all RBS spectra of oxidized samples. Chlorine present in the as-deposited layer [15], is not detected in the oxidized upper part of the sample: see enlarged part of the spectrum near channel 320.

From the RBS peak areas of N, O and (amorphous) Si the corresponding number of atoms per cm^2 was determined for each element following the procedure described before [16]. By comparison with the as-deposited layer, the increase in the number of O atoms due to oxidation is thus obtained and referred to as ΔO hereafter. These ΔO-values were corrected for spread in deposited oxynitride film thickness (± 5 %), which is inherent to the process of LPCVD applied [14]. This correction was accomplished by multiplying the O and N contents found for the as-deposited sample by a factor equal to the ratio of the Si content of the oxidized sample relative to that of the as-deposited film.

We verified that, for each sample, the ΔO-value corresponded reasonably well to 3/2.ΔN-value, as should be the case when N and O atoms in these layers remain surrounded by 3 and 2 Si atoms, respectively.

The thickness of the oxide layer formed upon oxidation may be estimated from the ΔO-value by dividing through the atomic density of O in SiO$_2$ (4.64 . 10^{22} cm^{-3}). This result must be multiplied by (1+2/3*O/N) to account for the oxygen already present in the initial oxynitride. The effect of this factor is visualized by the two vertical scales in fig. 4.

ERD

When the oxygen contents are small (2.10^{16} cm^{-2} or less) the sensitivity in RBS limits the accuracy of the determination of ΔO. In ERD an interfering background signal is absent, which makes the results of this technique both more sensitive and reliable for monitoring the first stage of the oxidation process. ERD measurements were performed using a primary beam of 30 MeV ^{28}Si ions, a scattering angle of $35°$ and a 8.9 μm thick Mylar absorber foil. Further experimental details have been given before [15].

An example of an ERD spectrum of an oxidized nitride film is shown in fig. 2. Note that the two oxygen contributions, originating from the native oxide at the substrate interface and the oxide layer at the surface, are well resolved. As may be deduced from fig. 2, the N and O signals will start to overlap for oxynitride films thicker than 50 nm. For this reason ERD was applied only to films thinner than this value.

Just as for RBS, values of ΔO were derived from ERD spectra by determining the total amount of oxygen before and after oxidation. In general, the ERD and RBS results were found to agree well, the differences always being smaller than $2.10^{16}cm^{-2}$ for ΔO, and in most cases less than 5.10^{15} cm^{-2}.

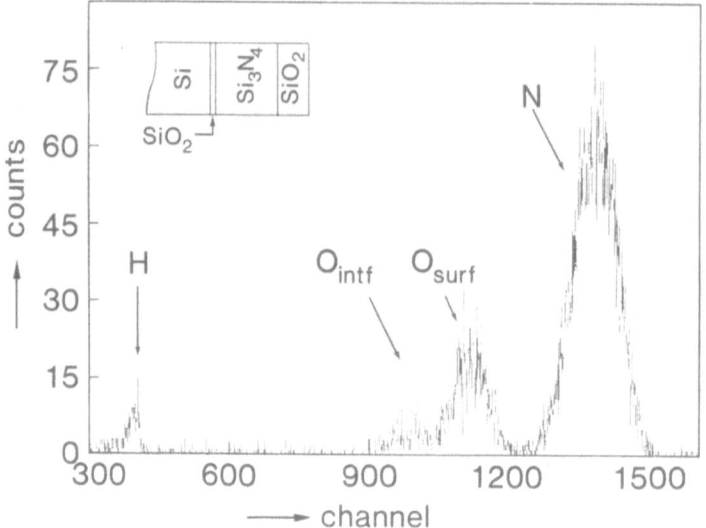

Fig. 2. ERD spectrum of 40 nm Si_3N_4 oxidized for 4 h in $H_2O/O_2 = 75/25$, 3 % HCl. The oxygen signal shows two distinct peaks: the larger one is due to the oxide layer grown, the smaller one results from the native oxide layer at the Si_3N_4/Si interface

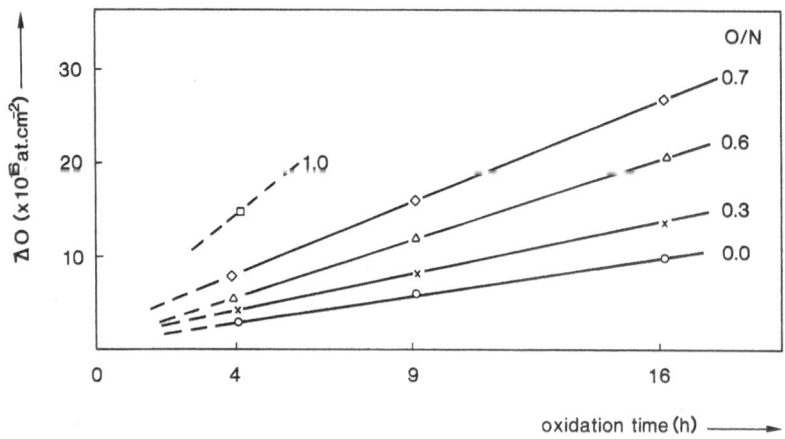

Fig. 3. Increase of the oxygen content as a function of time, measured in oxynitride layers of different compositions for wet oxidation at $1000^{0}C$ without HCl addition

II-3 Hydrogen analysis

Hydrogen contents and profiles were determined using NRA and ERD. With NRA hydrogen depth-profiles were determined in a selected number of samples employing the resonant nuclear reaction $^1H(^{15}N,\alpha\gamma)^{12}C$ at 6.39 MeV. Details of the experimental set-up have been published before [15]. This technique has a depth resolution of 5 nm at the surface, but is rather time-consuming when the hydrogen concentration in the sample is less than one atomic percent.

Therefore we also applied ERD, now with a primary beam of 2 MeV He ions. The scattered He ions and all recoiled particles except for the protons were stopped in a Mylar foil of 9 μm thickness. A depth resolution of about 10 nm is achieved with the incident beam at a grazing angle of 5° with respect to the sample surface [17]. The recoiled particles are detected at a scattering angle of 30°. These ERD measurements were performed in an UHV scattering chamber; during analysis the pressure was typically 1.10^{-8} Torr.

III. RESULTS

III-1 OXYGEN

III-1-1 Time dependence

Throughout this study the oxidation of various films is quantified by the increase in oxygen content, expressed in atoms per cm^2, relative to the corresponding unoxidized film. The plots in fig. 3 reveal how oxidation proceeds at $1000^{\circ}C$ for different oxynitride compositions with increasing reaction time. Straight curves are found for all O/N ratios, though they all cross the ΔO-scale at a positive value near $1.10^{16}at.cm^{-2}$. Presumably the oxidation of the first few atomic layers proceeds at an increased rate. For O/N = 1.02 the 100 nm-thick deposited layer was completely converted. The oxide thicknesses formed in 16h of wet oxidation range from 20 to 100 nm for O/N=0 and O/N=0.73, respectively. Comparison with an oxide thickness of 1900 nm grown on Si under the same conditions reveals the superior oxidation resistance of silicon oxynitrides.

III-1-2 Effect of HCl addition

    The oxidation rate of oxynitride films is increased when small amounts
of HCl are injected into the furnace. Fig. 4 shows the results obtained when

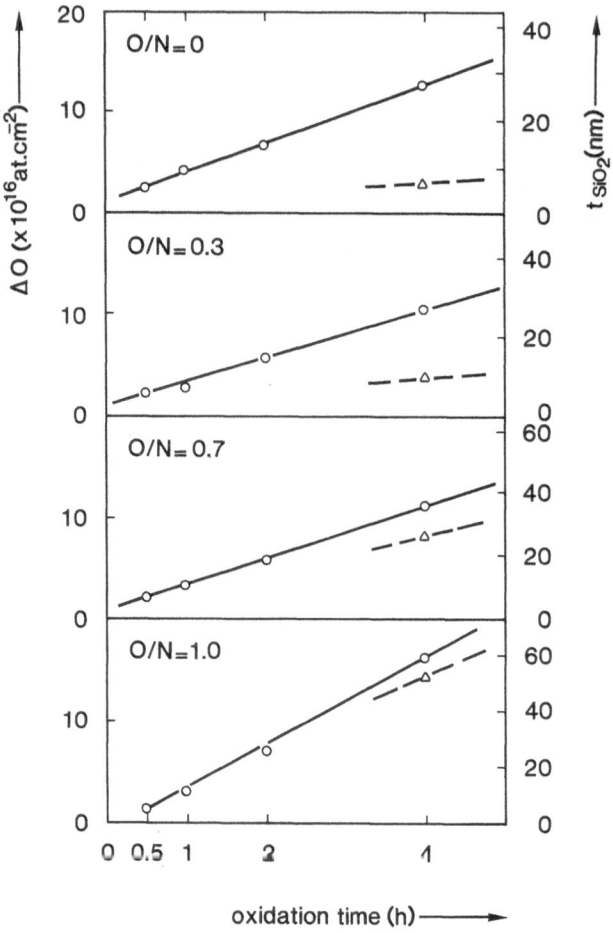

*Fig. 4. Wet oxidation at 1000°C of oxynitride layers in an ambient containing
3 vol % of HCl . For comparison corresponding data points of fig. 3 for HCl
free oxidation are included. The right hand scale indicates the grown oxide
thicknesses.*

3% HCl was added to the (wet) oxidizing ambient. For comparison, the data for

4h of fig. 3 (no HCl added) are included. During the first 4 hours of HCl-oxidation the data can be fitted reasonably well to straight curves. However, as reported previously the curves deviate from linearity for longer oxidation times [18]. For the lower O/N ratios the rate has appeared to slow down for longer times.

The effect of HCl is largest for pure Si_3N_4 and fades out with increasing O/N ratio. This is shown in fig. 5, where the rates, determined from the slopes of the curves in figs. 3 and 4, are compared. The acceleration induced

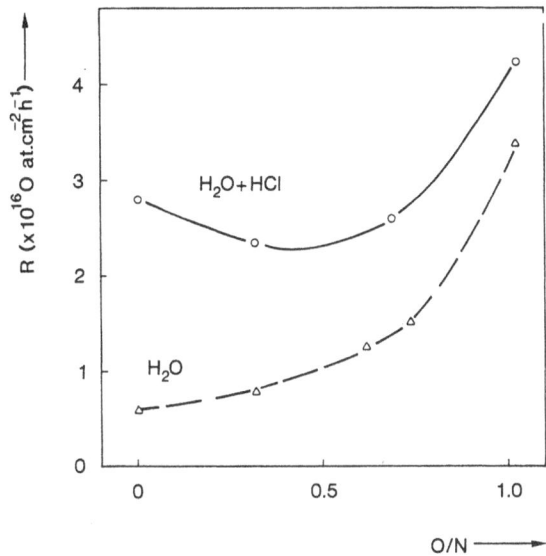

Fig. 5. Oxidation rates as determined from figs. 3 and 4 for different oxynitride compositions, revealing the accelerating effect of 3 % HCl in the oxidation ambient.

by HCl decreases from a factor of 5 for Si_3N_4 to 1.25 for O/N = 1.0. Fig. 5 also reveals that the earlier observed minimum in oxidation rate near O/N = 0.4 has shifted to O/N=0 when HCl is omitted.

Finally, we found no influence of the HCl concentration in the gas phase on the oxidation rate. We repeated the experiments for Si_3N_4 with HCl concentrations of 0.8% and 5% and found exactly the same results as with 3% HCl.

III-1-3 Effect of H_2O/O_2 ratio

Figure 6 gives the ΔO values after 4h oxidation at 1000°C as a function of H_2O/O_2 ratio in an ambient containing also 3 % HCl for various O/N ratios. It appears that the extent of oxidation increases more or less linearly with the water content of the ambient.

III-1-4 Effect of pre-annealing

In fig. 7 the oxidation behaviour of as-deposited films is compared with that of films annealed at $1000°C$ in N_2 prior to oxidation. This plot shows

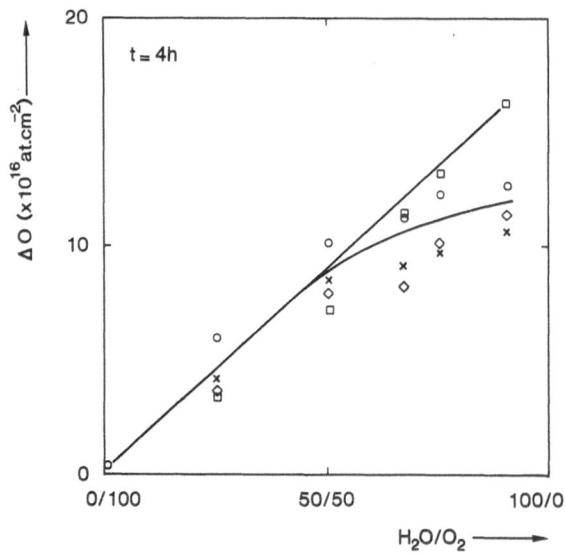

Fig. 6. Influence of of H_2O/O_2 flow ratio on the oxidation rate of various various oxynitride layers. Results plotted are for 1 hour oxidation at $1000°C$ and 3 % HCl added to the furnace.

that a preanneal has virtually no effect on the wet oxidation rate of silicon oxynitrides. In contrast, some reduction of the oxidation rate results in the case of HCl oxidation, but the effect is not the same for all compositions: the larger the O/N ratio the smaller the improvement in oxidation resistance.

III-1-5 Temperature dependence

The values for the oxygen up-take (ΔO) during oxidation of oxynitride films of about 50 nm thickness in an ambient with H_2O/O_2 =88/12 without HCl addition at T= 850 and 950°C are given in fig. 8 as a function of the oxidation time for as deposited as well as preannealed samples. It is clear that the extent of oxidation is smaller at 850°C than that at 950°C, which on its turn is smaller than that at 1000°C (cf. fig. 3). Within experimental error, the oxidation proceeds linearly with time after a relatively fast initial stage, which has also been observed in the 1000°C experiments.

The slopes of the linear parts of the curves of figs. 3 and 8 have been

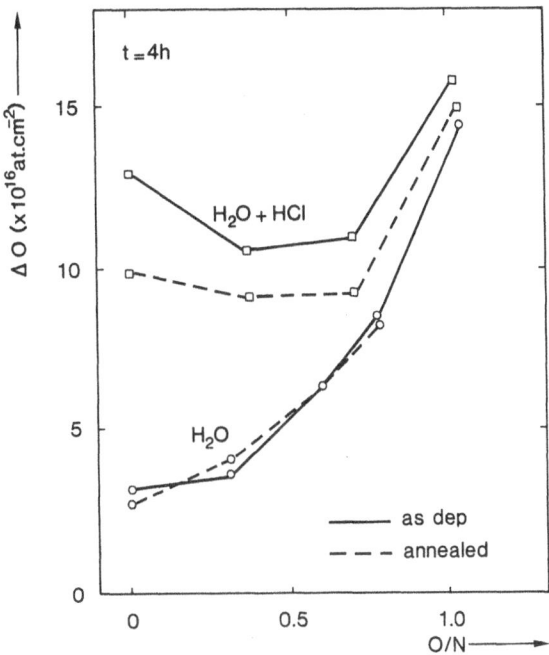

Fig. 7. Effect of preanneal on the wet oxidation of oxynitride films at 1000°C. Solid curves: as deposited films.; Dashed curves: annealed for 1 h at 1000°C in N_2 prior to oxidation.

plotted in an Arrhenius type diagram (fig. 9a) for the various O/N ratios. Within the experimental accuracy straight lines appear, the slopes of which are plotted in figure 9b as a function of O/N. It follows that the apparent activation energy for the linear stage of the oxidation process shows a tendency to increase at small O/N ratio but decreases with increasing O/N from ≃50 kcal/mole at O/N=0.32 down to 30 kcal/mole at O/N=0.76. Within the experimental uncertainty a 1000°C preanneal does not affect the value for the activation energy.

III-2 HYDROGEN

III-2-1 Wet oxidation

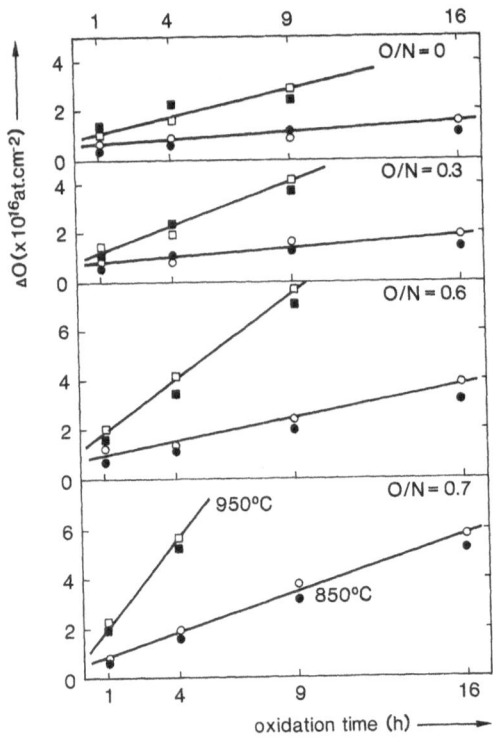

Fig. 8. Oxygen uptake during oxidation in a HCl free ambient with H_2O/O_2=90/10 at 850°C and 950°C at various oxynitride compositions.

NRA measurements of oxidized oxynitride layers reveal hydrogen profiles that can be conceived as consisting of three regions (fig. 10): i) a surface region, corresponding in thickness to the oxide grown, containing a very low level of H, ii) an H peak at the SiO_2/SiO_xN_y interface region, and iii) a

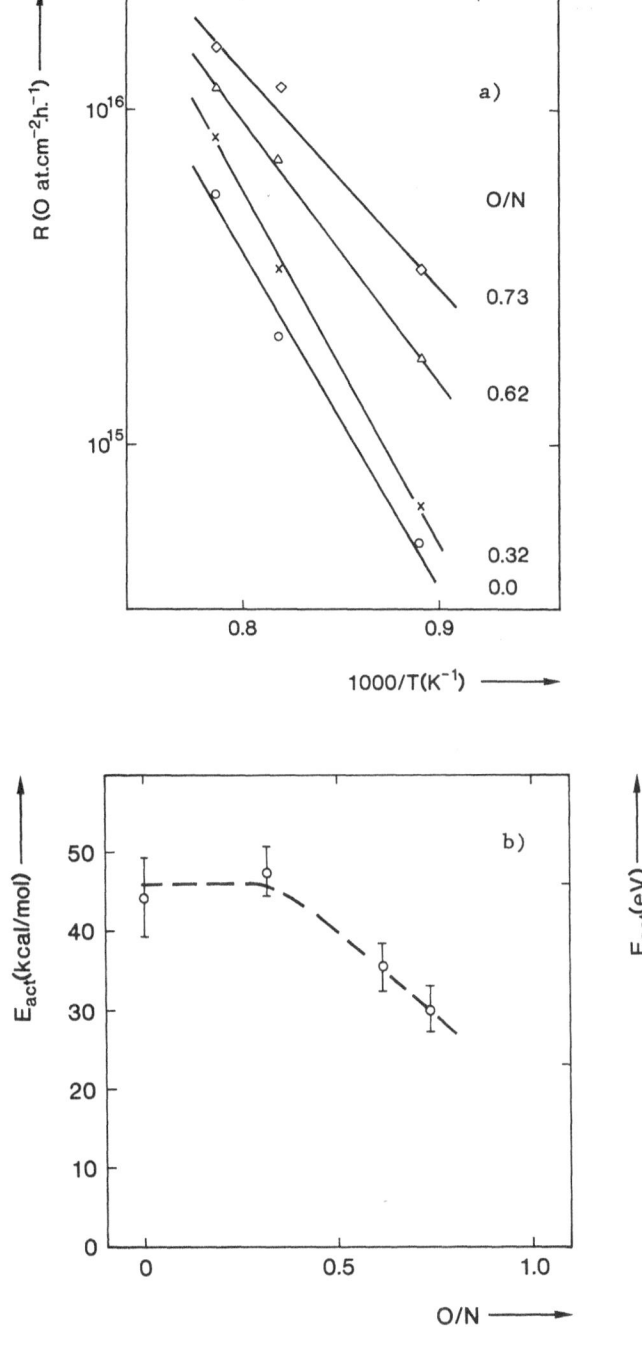

Fig. 9. *Arrhenius plot of the oxidation rate for various oxynitride compositions (a). The apparent activation energy for the oxidation of silicon oxynitrides as function of O/N (b). Oxidation without HCl.*

constant level of H in the remaining oxynitride film. The observation of the interface peak may be obscured in the case of a relatively high H level in the oxynitride underneath (fig. 11). We have found before [15] that annealing at 1000°C substantially reduces the H content of oxynitride films and that this effect is partly irreversible [19]. During a subsequent wet oxidation step only a limited amount of H will be reincorporated, so that a much lower level is found for the pre-annealed sample of fig. 11. In fact, this lower level is about the same as that found for an unannealed film oxidized at 1000°C. To a first approximation, the H concentration in region iii) is determined by the highest temperature the oxynitride has been subjected to.

Of more interest may be the behaviour of the interface peak in the oxidation process. Quantitative information was extracted from NRA profiles, such as shown in fig. 10. using an iterative computer simulation procedure. In this procedure the thickness of the regions i) and iii), their H concentrations and the width and content of the interface peak were the adjustable parameters. The program takes the energy straggling of the ^{15}N ion beam into account. The solid line in fig. 10 represents the best fit in that example.

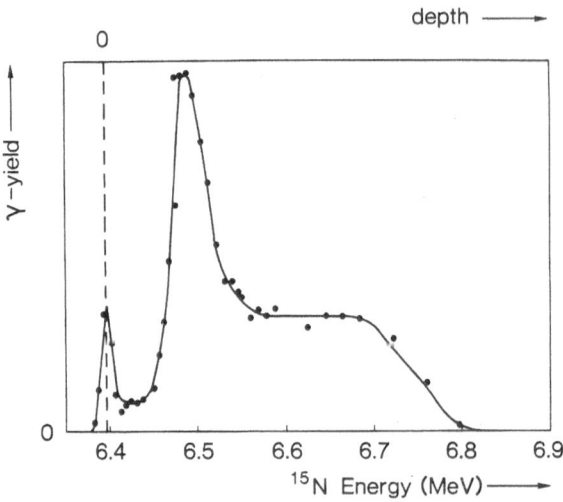

Fig. 10. NRA hydrogen profile in a 150 nm-thick oxynitride film with O/N=0.76 oxidized for 9 hrs at 950°C in H_2O/O_2=90/10 without HCl. Solid line represent the best fit of the profile, obtained from a simulation (see text).

The interface peak responds to process conditions such as oxidation time and oxynitride composition (see e.g. fig. 12). In Si_3N_4 samples the interface peak area was always measured to correspond to about 1.10^{15}at.cm^{-2}, whereas for O/N = 0.76 it varied from $2.6.10^{15}$ (t=1h, T = 850°C) to $7.5.10^{15}$at.cm^{-2} (t=9h, T=1000°C). As fig.12 shows, the interface peak may appear broadened when the interface is at larger depths. This is partly due to energy straggling of the recoiled protons, however, simulation shows that the H pile-up at the interface must extend over at least a 10 nm-thick region in the O/N = 0.76 sample.

The content of the interface hydrogen pile up increases with oxidation time in the early stage and reaches a saturation level after about 4h oxidation (fig. 13). This saturation level increases with increasing temperature. The width of the interface pile up behaves in a related manner. The local concentration of hydrogen at the oxide/oxynitride interface appears to amount to a few at.%.

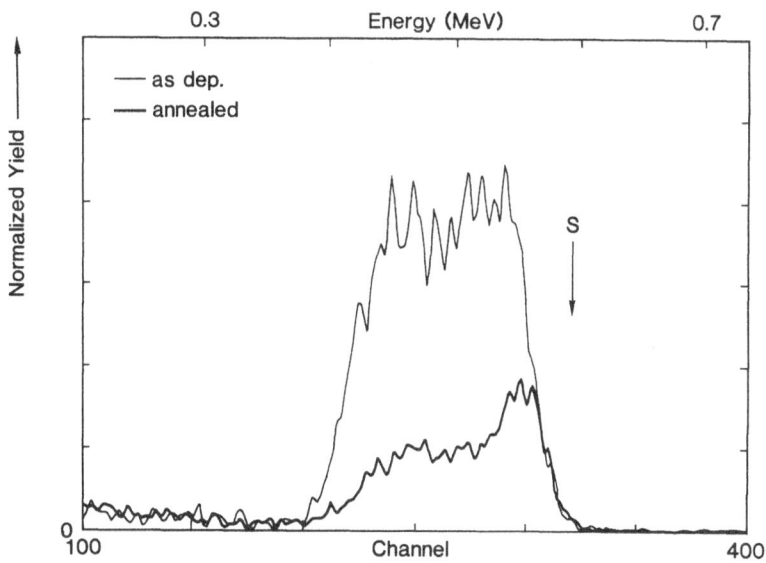

Fig. 11. He-ERD spectra of Si$_3$N$_4$ oxidized at 850°C for 4 h in a wet, HCl-free, ambient, showing the distribution of hydrogen in the samples. The energies corresponding to the surface and interface positions are indicated. In the spectrum for the sample that was annealed prior to oxidation all three regions mentioned in section III-2-1 are resolved.

III-2-2 Effect of HCl

Adding HCl to the oxidizing ambient affects the hydrogen profiles dramatically: fig. 14. Both the interface peak and the H level in the underlying (oxy)nitride layer are found to be reduced in the HCl-case. The interface peak is found at lower energy in the ERD spectrum, hence the hydrogen accumulation is present at larger depth, in agreement with the larger oxide thickness grown in HCl. In fig. 5 we showed that the effect of HCl on the oxidation rate diminishes towards higher O/N ratios. This does not apply to the hydrogen profiles. Instead, the difference in the H pile up between the HCl oxidized samples and the samples oxidized without HCl in the

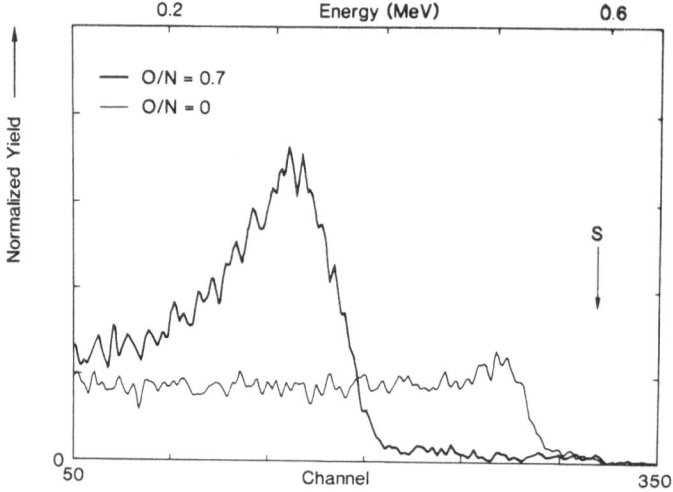

Fig. 12. He-ERD spectra measured after 16 h of wet oxidation at 1000°C (no HCl added) of 150nm-thick oxynitride films. Compared are the hydrogen distributions for two oxynitride compositions: O/N=0 and O/N=0.76.

ambient increases with O/N: fig. 15.

Neither the interface peak nor the level in the remaining oxynitride appeared to vary with the HCl concentration in the gas phase (0.8-5 vol.%). In this respect H behaves similarly to the oxidation rate (section III-1-2).

Since the H content in the oxynitride film after oxidation is much lower than after HCl-free oxidation, the interface peak can be quantified more accurately. In fig. 16 data are collected to demonstrate how H accumulates at the interface during oxidation of various oxynitride samples. The curves show that H piles up rapidly in the first hour of the oxidation process, after which a saturation level is reached, similar to the case of HCl-free oxidation. This level increases more than proportional with O/N ratio. The size of the interface peak is proportional to the humidity of the oxidizing ambient: fig. 17.

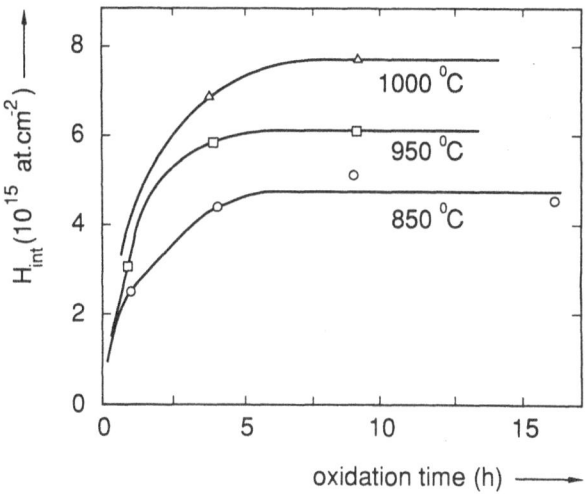

Fig. 13. Time evolution of the content of the hydrogen interface pile up during wet oxidation of the oxynitride with O/N= 0.76 for different temperatures.

Several HCl-oxidized samples were also measured using NRA. Figure 18 collects H profiles in Si_3N_4 for various oxidation periods. The interfacial H pile up is seen to shift inwards with ongoing oxidation time, consistent with a growing oxide layer. In addition a surface H peak is observed presumably due to adsorbed water or hydrocarbons (the pressure in the NRA vacuum system was in the 10^{-6} Torr range.)

Simulations of the profiles (solid curves in fig. 18) reveal that the content of the interface peak increases from $1.8.10^{14}$ at.cm^{-2} for t=0.5 h to 3.10^{14} H at.cm^{-2} for t=4h (H_2O/O_2 = 90/10), which is in good agreement with the ERD data in fig. 16. In addition, for O/N<0.5 the measured interface peak can be simulated assuming a monolayer-thick H-containing interface. For larger O/N ratios the width of the peak is somewhat larger, but in all cases it is much smaller than found in HCl free oxidized samples.

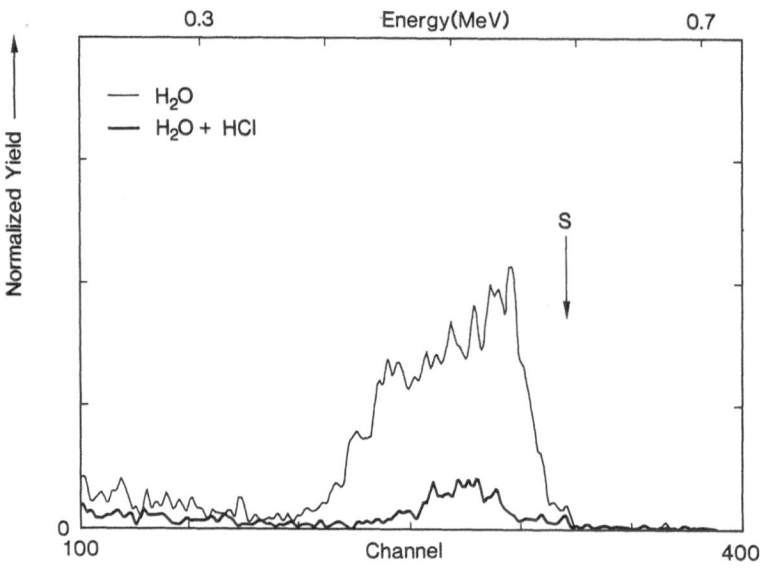

Fig. 14. Hydrogen profiles as determined using He ERD in 50 nm thick Si_3N_4 after 4 h of wet oxidation at $1000^{\circ}C$, with and without HCl addition in de ambient. The hydrogen concentration in the oxidized sample is much lower when HCl was added to the furnace.

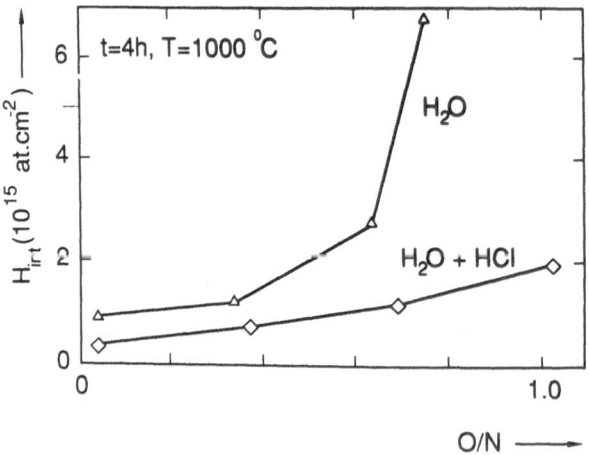

Fig. 15. O/N dependence of the amount of interfacial hydrogen after wet oxidation of oxynitride films for 4h at $1000^{\circ}C$ in an ambient with and without HCl.

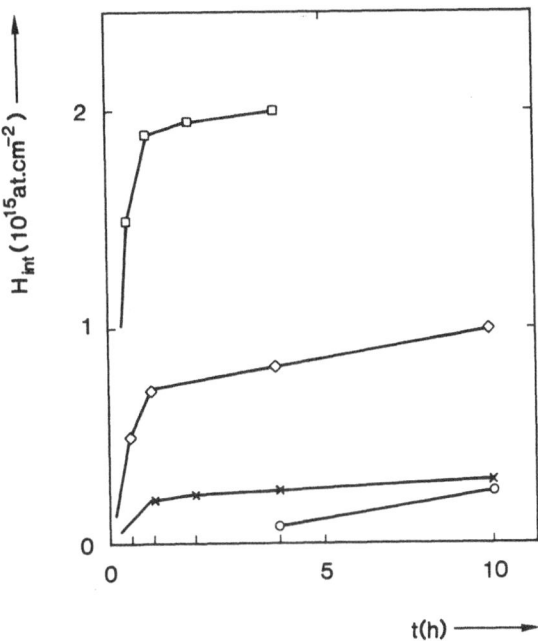

Fig. 16. Development of the hydrogen interface peak with oxidation time for the various oxynitride compositions; H_2O/O_2=75/25, 3 vol. % HCl added to the oxidation ambient. T=1000°C (o: O/N=0, x: O/N=0.37, :O/N=0.69, □:O/N=1.)

IV DISCUSSION

IV-1 Interface Reactions

In this study abundant proof was collected for hydrogen playing a role in the oxidation of silicon (oxy)nitride. Water not only accelerates the oxidation, it is probably a prerequisite for the reaction to occur. As we have reported recently [18] hardly any oxidation occurs n dry oxygen. These observations suggest that when hydrogen is absent no oxidation will occur. In dry oxidation the only source of hydrogen is the oxynitride layer itself. In the first 10 minutes of oxidation most of this hydrogen will desorb, as we have measured earlier [19]. Some of this hydrogen will contribute to oxidation, giving rise to a slight surface oxidation. This effect may account

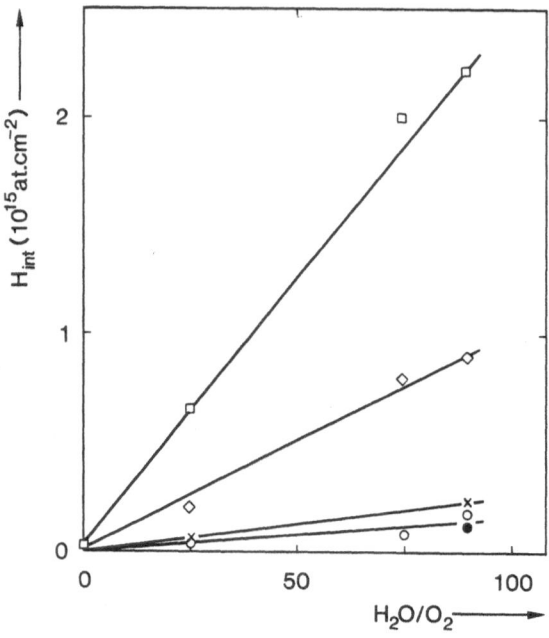

Fig. 17. Dependence of the size of the interface hydrogen peak on the water content in the oxidizing ambient. Results given are for t=4h and T=1000°C.

for the initially larger rate, as we deduced from extrapolation of the curves in figs. 3 and 8 to t= 0h. If no additional hydrogen is supplied from the ambient, as in the case of dry oxidation, the oxidation of the layer will terminate after this initial period.

Further support for the peculiar role of hydrogen in the oxidation process follows from the intriguing observation of a hydrogen peak at or near the oxidizing interface. We relate this peak to the oxidation process itself. As is illustrated by figs. 12-17, the peak size responds to variations in oxidation conditions, which precludes that this phenomenon is due to an artifact of the applied analytical techniques or to alterations in sample composition occurring after completion of the oxidation treatment.

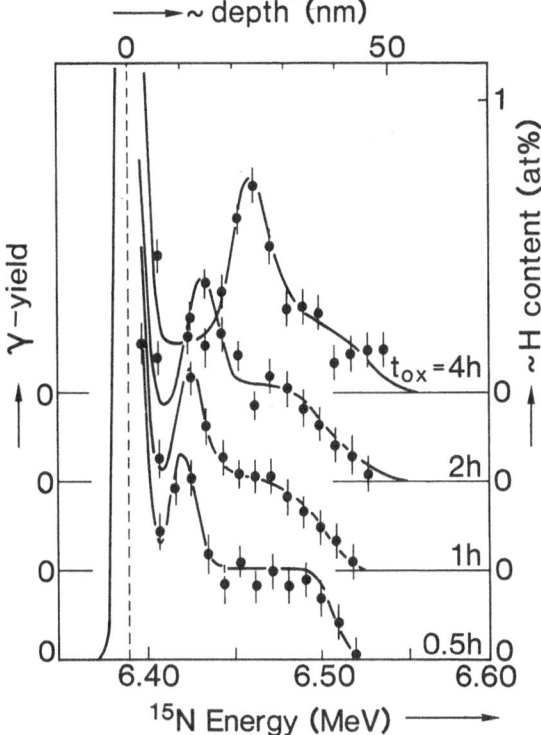

Fig. 18. NRA hydrogen profiles measured in Si$_3$N$_4$ films oxidized in an ambient of H$_2$O/O$_2$=90/10 with 3 vol % HCl added. The solid curves are computer simulations. The shift of the interface peak to larger depths with ongoing oxidation time corresponds to the growth of the oxide formed.

During oxidation Si-OH and Si-NH groups may be formed at the interface as intermediate reaction products. This is illustrated by the following reaction scheme:

$$\text{Si-N}\begin{array}{l}\diagup\,\text{Si}\\[-0.3em]\diagdown\,\text{Si}\end{array} + \text{OH}_2 \;\rightleftharpoons\; \text{Si-N}\begin{array}{l}\diagup\,\text{Si}\\[-0.3em]\diagdown\,\text{H}\end{array} + \text{Si-O-H} \qquad\qquad (1a)$$

$$\text{Si-N}\begin{array}{l}\diagup\,\text{Si}\\[-0.3em]\diagdown\,\text{H}\end{array} + \text{OH}_2 \;\rightleftharpoons\; \text{Si-N}\begin{array}{l}\diagup\,\text{H}\\[-0.3em]\diagdown\,\text{H}\end{array} + \text{Si-O-H} \qquad\qquad (1b)$$

$$Si-N\begin{matrix} H \\ \\ H \end{matrix} + OH_2 \underset{\longleftarrow}{\longrightarrow} NH_3 + Si-O-H \tag{1c}$$

SiO_2 may be formed via:

$$2\ Si-O-H \longrightarrow Si-O-Si + H_2O \tag{1d}$$

In this scheme the rupture of an Si-N bond is achieved through hydrogen bonding, possibly forming an $=N--H_2O--Si=$ transition complex. This would explain why no oxidation occurs in dry oxygen. Moreover, it has beenobserved that NH_3 is an end-product of the wet oxidation of oxynitrides [20, 21], which is in support of the above reaction scheme. In fact, the reaction mechanism proposed is the reverse of the mechanism that has been assumed to be operative in thermal nitridation of SiO_2 [22,23]. Hydrogen is essential for nitridation too: the reaction proceeds in NH_3, but not in N_2. Even N implantation in SiO_2 is not effective: N will be expelled from the oxide during a post-implantation anneal [24].

We assume that the measured H pile ups are present during the oxidation treatment itself. This is not obvious because it is conceivable that immediately after the oxidation treatment i.e. during the transport of the wafers out of the hot zone of the furnace under flowing N_2, part of the hydrogen desorbs or becomes trapped at the considered interface. Desorption of any physically dissolved hydrogen or hydrogen containing species from the material will probably occur during cool down after the oxidation [19,25], so the interface pile up has to be due to bonded hydrogen.

The observed proportionality of the amount of interface hydrogen with the humidity of the oxidation ambient can be understood from the reaction sequence 1a-1c since increasingly more water molecules are available to form the intermediates. The O/N dependence of the H interface pile up is more difficult to interpret. From the broadening of the interfacial hydrogenated region it may be deduced that the reaction volume increases upon an increase in O/N ratio.

According to the reactions 1a, b, c incorporation of H is accompanied by incorporation of O. H may also be incorporated via the inverse reaction 1d. In the top SiO_2 film this amount will be low since the solubility limit of H_2O in SiO_2 is in the order of a few tenths of an atomic percent [25]. In fact the values for the H concentration in the oxide-part of the layer structures are in good agreement with the solubility limit reported in ref. 25.

The variation of the local hydrogen concentration and the variation in hydrogen interface width with O/N ratio may be discussed as follows: Hydrogen is incorporated at the reaction interface via reactions 1a, b and c. This amount will decrease with a decreasing number of N atoms which can be reached by H_2O molecules. So the inserted amount of hydrogen is determined by the local N concentration and the permeability for water molecules of the oxide/oxynitride interface. The latter phenomenon will also affect the interface width. On he other hand, hydrogen is removed from the reaction interface through reaction 1d. A lower concentration of nitrogen increases the average distance between two Si-OH groups which may result in a lower removal rate of hydrogen. The net effect may be an increasing hydrogen concentration as well as in an increasing hydrogen interface width with increasing O/N.

IV-2 Oxidation kinetics

IV-2-1. Rate limiting step

The oxidation process of a silicon oxynitride film may be conceived as a three-step mechanism: 1) transport of oxidant through the grown oxide to the oxidizing reaction front, 2) the actual oxidation reaction and 3) transport of reaction products back to the solid-gas interface. The oxidation reaction may be represented as:

$$OH_x + SiON \longrightarrow SiO_2 + NH_x \qquad (2)$$

which is a simplification of the reaction equation (1a)-(1d).

It was mentioned in section III.1 that the oxidation rates of oxynitrides are much lower than that of Si or SiO_2. Since, to our knowledge, no evidence exists which indicates that the oxide grown on silicon is a material widely different from that formed by oxidation of (oxy)nitride, step (1) of the process will be indistinguishable from OH_x diffusion through a conventional thermal oxide and can, therefore, be precluded as being rate-limiting in the case of oxynitride oxidation.

In view of the larger bond strength of Si-N as compared to Si-Si, the reaction rate in the second step may very well be lower than in the case of

Si oxidation. As a result, the oxidation process may remain in its reaction-controlled regime for considerably larger oxide thicknesses than observed for Si oxidation, and would show linear kinetics for longer reaction times and thicker grown oxides. The plots in figs. 3 and 8 indeed reveal a linear dependence of oxygen incorporation on time, for all oxynitride compositions investigated.

Similar behaviour is displayed in fig. 4 for the HCl case. However, we have found earlier [18] that for reaction times greater than 4h the curves deviate from linearity, but not in a similar way for all O/N ratios. The oxidation rate gradually slows down in the case of a low O/N ratio, but speeds up for O/N = 1. We assign this transition to non-linear kinetics to the outdiffusion of reaction products becoming rate limiting (step (3)), although it is also feasible that different reactions become involved.

So, for the conditions pertinent to figs. 3 and 4 we propose that the oxidation of silicon nitride and oxynitride layers is reaction controlled. Hence, if equation (2) is an adequate representation of the interface reaction, the oxidation rate may be expressed as:

$$v = \frac{[OH_x]}{[NH_x]} \cdot A \cdot e^{-\frac{E}{RT}} \tag{3}$$

where A is the frequency factor and E the activation energy as in an Arrhenius equation. In addition we suppose that HCl stimulates the removal of NH_x species from the reaction interface; how this may be accomplished will be discussed below.

With this model, various observations can be explained directly. For instance, adding HCl to the ambient will reduce $[NH_x]$ at the reaction interface. This will affect the equilibrium in equation (2) and result in a larger oxidation rate v, whereas the activation energy E remains the same. Thus, a larger rate causes a net reduction of $[OH_x]+[NH_x]$ at the interface, which is in agreement with the smaller interface H peaks found for the HCl-oxidized samples. Further, reaction (2) is first order in $[OH_x]$, so the oxidation rate is expected to depend linearly on the H_2O/O_2 ratio in the gas phase. This was indeed found to be the case [18], as is shown in fig. 6: the rate varies linearly with H_2O/O_2 up to a ratio of 50/50, above which some saturation occurs for the lower O/N samples. Moreover, fig. 17 shows that the interface H peak is also linearly proportional to this H_2O/O_2 ratio, which we take as being in support of expression (3).

IV-2-2. Effect of O/N

Figs. 5 and 9a show that the oxidation rate, expressed as the number of oxygen atoms taken up per unit time, increases, in the considered temperature region 850-1000°C, with O/N ratio of the oxynitride film notably for O/N>0.5. The measured activation energy for the oxidation process decreases in that O/N region. These effects cannot be explained on the basis of the reactions 1a-d alone, since oxygen of the oxynitride is not considered in that reaction scheme. Therefore, a non-local phenomenon must be responsible for the O/N dependence of the oxidation rate and its activation energy. A candidate for this non-local effect might be the structural transition from (oxy)nitride to silicon dioxide. In this view the measured activation energy is not singly the activation energy for the Si-N bond breaking under simultaneous oxygen insertion but it also comprises a contribution of the rearrangement of the network. At first glance the only network modifying reaction is reaction 1d. The occurrence of this reaction is necessary to shift the reactions 1a, b and c to the right hand side via the removal of Si-OH groups. Within this view reaction 1d is identified as the rate limiting reaction. In this respect it is interesting to note that the values for the activation energy of the oxynitrides with O/N<0.32 are close to those observed for the linear rate constant in the steam oxidation of Si [26]. This linear rate constant is supposed to be associated with the interface reaction in the Deal-Grove description of the oxidation of silicon [27].

The decrease in activation energy for O/N>0.32 indicates that the oxynitride network is increasingly more "oxide like" whereas the constant activation energy for O/N<0.32 indicates that the network is "nitride-like". Such a structural transition in the as deposited LPCVD oxynitrides has also been discussed in chapter I of this book.

IV-2-3. Effect of HCl

Although the presence of HCl accelerates the oxidation to a large extent, its presence in an otherwise hydrogen free ambient does not result in an appreciable oxidation rate.

The incremental oxidation rate due to HCl diminishes for larger O/N: compare the two curves in fig. 5. In order to interpret this effect one has to speculate about the way HCl is involved in the oxidation mechanism. As stated in section IV-2-2 the rate limiting step of the oxidation might be the combination of two Si-OH groups under release of H_2O accompanied by a rearrangement of the network. The role of HCl may, in this view, be conceived as to accelerate this rearrangement. The structural rearrangement, which is necessary to form SiO_2, is larger in the nitrogen rich oxynitrides and therefore the effect of HCl is larger in that O/N range. Consistently, the presence of HCl is responsible for a smaller amount of interfacial hydrogen.

Another explanation for the role of HCl might be that its presence accelerates the transport of NH_x out of the sample. This possibility is based on the observation that HCl increases the oxidation rate without affecting the activation energy E. Hence, the rate limiting reaction is the same in either case.

The action of HCl may be conceived in various ways. Chlorine may be incorporated in the growing oxide, thus making this layer more permeable to NH_x and OH_x. This will enhance $[OH_x]$ and will reduce $[NH_x]$, so that the rate will increase according to equation (3). However, using RBS and Auger electron spectroscopy we could not measure any Cl in the oxide and, moreover, chlorine incorporation would not explain the observed dependence on O/N.

A more plausible explanation is that HCl adsorbs at the surface of the oxide and reacts with arriving NH_y species to form a readily desorbing compound, e.g. NH_4Cl. (Note, that NH_4Cl is a reaction product in the LPCVD of Si_3N_4 using SiH_2Cl_2 and NH_3). In this way the concentration gradient of NH_3 over the oxide layer will be increased which will lead to a lower $[NH_x]$ at the interface. This effect will be largest at the higher N contents, hence for the lower O/N ratios. Alternatively, the adsorption of Cl may induce (or increase) an electric field across the dielectric bilayer such that the NH_y diffusion is enhanced, similar to the mechanism proposed recently by Wolters and Zegers for Si oxidation [28]. This field effect will depend, among others, on the dielectric constants involved and thus depend on O/N.

The ERD spectra, like those in fig. 14, show that the effect is not limited to the interface but extends to the remaining oxynitride layer underneath, where the hydrogen content is also found to be reduced relative to the hydrogen-free situation. We see no reason why the diffusion enhancement of HCl should be restricted to NH_y: it also may affect hydrogen desorption.

We verified this by annealing oxynitride samples in dry oxygen (so that no oxidation will occur) with and without 3 % HCl. Fig. 19 compares He ERD

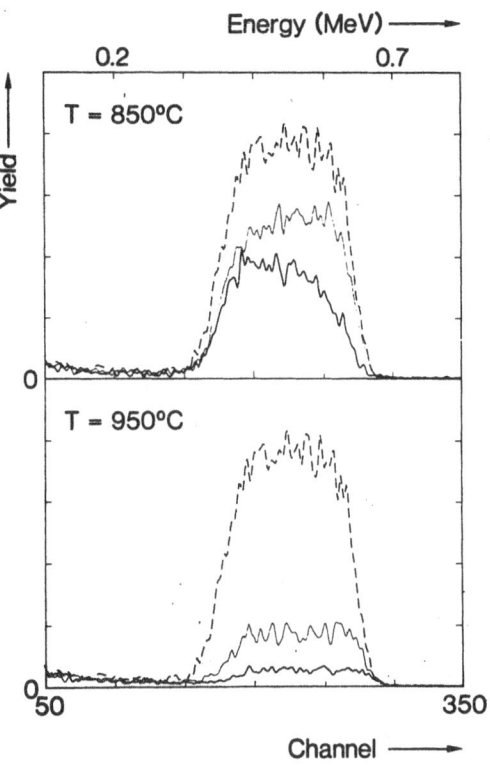

Fig. 19. He ERD spectra of 80 nm thick Si_3N_4 films. Dashed: as deposited, continuous curves: after 1 h annealing at the temperature indicated; ——— in O_2, —— in O_2 + 3 % HCl.

spectra of a few of these samples, proving indeed that HCl accelerates the desorption of H from Si_3N_4. Note the gradient in the H profile near the surface of the sample annealed at 850°C in O_2+HCl. With RBS we established that, as expected, oxidation of these samples– if it occurred at all– was less than 10^{16} O at/cm^2 (less than 2 nm SiO_2 formed) at 950°C. Fig. 19 presents results for Si_3N_4 only, but the same effect was observed for an oxynitride sample with O/N=0.7, although the difference between annealing in O_2 and in O_2+HCl was smaller in this case. Note that the enhancement of the oxidation rate induced by the presence of HCl displays a similar O/N dependence: see fig. 5. Although the results given in fig. 19 do not provide information about the nature of the interaction of HCl with oxidizing oxynitrides, they strongly support the concept of increased outdiffusion.

Franz and Langheinrich [6] have observed that adding PBr_3 to the oxidation furnace increases the oxidation rate of APCVD Si_3N_4 (and found linear kinetics in that case!). This was attributed to incorporation of P in the oxide, but it is also feasible that the adsorption of Br is responsible for the effect.

The minimum in oxidation rate observed in HCl assisted oxidation near O/N=0.4 (see fig. 5) can be understood to result from two additive processes, which depend oppositely on O/N: the decreasing activation energy with O/N, causing the rate to increase with O/N, and superimposed on this the accelerating effect of HCl which diminishes with increasing O/N. The resulting minimum has - in this picture- no deeper physical background.

IV-2-4. Effect of preanneal

The plots in figs. 7 and 8 demonstrate that a preanneal has only a very limited effect on the oxidation process when no HCl is added. Annealing strongly reduces the H content of the material [15], but apparently the hydrogen present in the as deposited material plays no active role in the oxidation.

In the HCl-assisted process the effect of the anneal treatment is larger and O/N-dependent: fig. 7 indicates a difference in ΔO of 3.10^{16} at/cm^2 at O/N=0 after 4h of wet oxidation at 1000°C. In view of the ideas outlined in section IV-2-3, it is conceivable that notably during the initial stage of HCl assisted oxidation the H desorbing from the bulk of the (oxy)nitride gets involved in the oxidation process, for example by recombination with adsorbed HCl and arriving NH_y species at the surface, forming NH_4Cl. Indeed, for larger O/N ratios, with lower H contents in the as-deposited material [15], the effect of pre-annealing becomes smaller (fig. 7.). However, without the specific knowledge of the role of HCl in the oxidation process any interpretation of the pre-anneal effect will remain rather speculative.

V. CONCLUSIONS

LPCVD silicon nitride and silicon oxynitride (O/N≤1) were measured to oxidize in a wet ambient at rates that are nearly two orders of magnitude smaller than that found for Si oxidation under identical conditions. After a rapid initial oxidation of the surface region the process displays linear kinetics in the time and temperature region of this study (up to 16 h, 850 -

$1000°C$). This observation identifies the oxidation of (oxy)nitride as a reaction-controlled process.

The value for the apparent activation energy of the oxidation process in the considered temperature range is 45-50 kcal/mole for O/N<0.4 and decreases for larger O/N rations to 30 kcal/mole at O/N=1.

A model for the actual oxidation reactions proceeding at the interface is proposed, in which the bonds between a single N atom and its three neighbouring Si atoms are consecutively broken and O is initially inserted as an OH group. In this scheme hydrogen bonding is essential for Si-N bond-breaking to occur, which would explain why (oxy)nitride does not oxidize in dry oxygen.

We found evidence for an accumulation of hydrogen species during oxidation near or at the oxidizing interface. This interface peak reaches a saturation level after 1-2 h oxidation at $1000°C$. The extent of accumulation increases with O/N ratio and water content of the ambient. In accordance with the model this interface peak is assigned to NH and OH species present at the reactive interface, which at least allows us to interpret all H-related results in a consistent manner.

When HCl is added to the furnace the oxidation rate of nitride is found to be enhanced by roughly a factor of 5, but this accelerating effect declines with increasing O/N ratio. One of the proposed models states that HCl activates the transport of NH_y species from the reaction interface to the surface. Indeed, less H is found in HCl oxidized samples, and in particular the interface peak is smaller and narrower than in the case of HCl-free oxidized oxynitrides. It has been established that the presence of HCl in the gas phase accelerates the desorption of H from silicon oxynitrides under conditions where no significant oxidation takes place. For oxidation periods exceeding t=4h the HCl-assisted process starts to deviate from linearity. However, on the basis of the results collected in this study it is not possible to assess this phenomenon unambiguously.

Ackowledgments

The authors would like to thank Henry Cox, Peter Snee and Michael Nieuwesteeg for sample preparation and the technical staff of the accelerator center of Utrecht University for running the tandem accelerator.

REFERENCES

[1]. J.A. Appels, E. Kooi, M.M. Paffen, J.J.H. Schatorjé and W.H.C.G. Verkuylen, Philips Res. Rep. 25,118 (1970).

[2]. P.A. van der Plas, W.C.E. Snels, A. Stolmeijer, H.J. den Blanken and R. de Werdt, *IEEE-EDS 1987 Symposium on VLSI Technology*, Karuizawa, Japan (1987)

[3]. E.A. Lewis and E.A. Irene, J. Vac. Sci. Technol. A4, 916 (1986).

[4]. D.R. Wolters, J. Electrochem. Soc. 127, 2072 (1980)

[5]. I. Fränz and W. Langheinrich, Solid State Electr. 14, 499 (1971).

[6]. T. Enomoto, R. Ando, H. Morita and H. Nakayama, Jap. J. Appl. Phys. 17, 1049 (1978).

[7]. L.V. Chramova, T.P. Smirnova, B.M. Ayupov and V.I. Belyi, Thin Solid Films 78, 303 (1981).

[8]. M.E. Washburn, Am. Ceram.Soc.Bull. 46, 667 (1967).

[9]. I.Ya Guzman, Yu.N. Litvin and G.V. Turchina, Ogneupory 2, 47 (1974).

[10]. P. Goursat, P. Lortholary, D. Tetard and M. Billy, *Proc. 7th Int. Symp. Reactivity of Solids (Bristol)*, Ed. J. Anderson et al. (Chapman Hall, London, 1972), p. 315.

[11]. O.J. Gregory and M.H. Richman, J. Am. Ceram. Soc. 67, 335 (1984).

[12]. S.C. Singhal, J. Mater. Sci. 11, 500 (1976).

[13]. J.G. Dil, J.W. Bartsen, R.D.J. Verhaar and A.E.T. Kuiper, Philips J Res. 40, 72 (1985).

[14]. A.E.T. Kuiper, S.W. Koo, F.H.P.M. Habraken and Y. Tamminga, J. Vac. Sci Technol. B1, 62 (1983).

[15]. F.H.P.M. Habraken, R.H.G. Tijhaar, W.F. van der Weg, A.E.T. Kuiper and M.F.C. Willemsen, J. Appl. Phys. 59, 447 (1986).

[16]. F.H.P.M. Habraken, A.E.T. Kuiper, A. van Oostrom, Y. Tamminga and J.B. Theeten, J. Appl. Phys. 53, 404 (1982).

[17]. A.E.T. Kuiper, M.F.C. Willemsen and L.J. IJzendoorn, Apl.Phys. Lett., to be published.

[18]. A.E.T. Kuiper, M.F.C. Willemsen, J.M.G. Bax and F.H.P.M. Habraken, Appl. Surf. Sci.33/34 (1988) 757.

[19]. J. B. Oude Elferink, Thesis University of Utrecht, 1989. To be published.

[20]. S. Kainaskii, E.V. Degtyareva and V.A. Kuhtenko, Ogneupory 25, 175 (1960).

[21]. C. Contet, J.I. Kase, T. Noura, M. Yoshimura, S. Somiya, J. Mat. Sci. Lett. 6, 963 (1987).

[22]. F.H.P.M. Habraken, A.E.T. Kuiper, Y. Tamminga and J.B. Theeten, J. Appl. Phys. 53, 6996 (1982).

[23]. A.E.T. Kuiper, M.F.C. Willemsen, A.M.L. Theunissen, W.M. van de Wijgert, F.H.P.M. Habraken, R.H.G. Tijhaar, W.F. van der Weg and J.T. Chen, J. Appl. Phys. 59, 2765 (1986).

[24]. W.J.M.J. Josquin and Y. Tamminga, J. Electrochem. Soc. 129, 1803 (1982).

[25] . G.H.A.M. van der Steen and E. Papanikolau, Philips Res. Repts 30, 192 (1975).

[26]. R.R. Razouk, L.N. Lie and B. E. Deal, J. Electrochem. Soc. 128, 2214 (1981).

[27]. B.E. Deal and A.S. Grove, J. Appl. Phys. 36, 3770 (1965).

[28]. D.R. Wolters and A.T.A. Zegers-van Duynhoven, J. Appl. Phys. (submitted).

Chapter 4

ELECTRICAL PROPERTIES OF LPCVD SILICON-OXYNITRIDE LAYERS

M. Heyns, J. Remmerie, E. Dooms, H. Maes and R. De Keersmaecker

Interuniversity Microelectronics Center (IMEC)
Kapeldreef 75
B-3030 Leuven, Belgium

ABSTRACT : *LPCVD silicon-oxynitrides with compositions ranging from pure oxide to nitride are examined on their electrical properties. The in-depth charge distribution of the layers, the interface state density and the retention reveal a transition at on O/(O+N) ratio of approximately 0.3 to 0.4. Other characteristics, such as the breakdown strength, the barrier height and the conduction through the layer show a more continuous behaviour as a function of the O/(O+N) ratio.*

1. INTRODUCTION

Silicon-oxynitrides deposited by LPCVD are a promising material for various applications in a VLSI-technology because it can be expected that the good characteristics of SiO_2 and Si_3N_4 can be combined for some compositions. In this chapter the electrical characteristics of silicon-oxynitrides deposited by Low Pressure Chemical Vapour Deposition (LPCVD) are explored as a function of the composition which is varied from oxide to nitride. The electrical characteristics which are assessed include the charge distribution in the layers, the interface trap density at the oxynitride/Si interface, the conduction and trapping in the layers, the retention and the dielectric integrity.

2. SAMPLE PREPARATION

The results reported in this chapter were obtained on samples which were prepared using the general deposition parameters described in chapter 1. The oxynitride depositions were carried out in a TEMPRESS LPCVD nitride furnace at a temperature of 800°C at the center zone (front zone 790°C and end zone 810°C). On some wafers a thermal oxide was grown prior to the oxynitride deposition. Oxides with a thickness of 3 nm were grown in a dry O_2 ambient at 700°C after an RCA pre-clean. 30 nm oxide layers were grown at 900°C, also in dry O_2.

The effect of various anneal treatments on the characteristics of the oxynitride layers was investigated. The anneal sequence which was used is summarized below.

- As-deposited

    - N_2 anneal, 1000°C, 1 hour

        + H_2 anneal, 1000 °C, 1 hour

            + plasma-H anneal, 300°C, 1 hour
              (100 sccm H, 123 mTorr, 300 W, 13.57 MHz, parallel plate reactor)

    - NH_3 anneal, 900°C, 1 hour

The H_2 anneal was performed on samples which previously underwent the N_2 anneal. The samples were annealed in an epi-reactor under an hydrogen flow of 85 sl/min. The plasma-H anneal was performed after the H_2 anneal in a Plasma Technology PE80 parallel plate system.

Aluminium gated capacitor structures were fabricated using standard lithography and dry etching after sputtering of Al/1%Si. Capacitor areas ranged from 10^{-5} cm^2 to 10^{-2} cm^2. On all samples a sintering was performed at 435°C for 30 min.

3. CHARGE DISTRIBUTION IN THE OXYNITRIDE LAYER

The *total charge* present in the oxynitride layers is an important parameter when these layers are to be used as dielectric layers. This charge is directly related to the flatband voltage of metal-oxynitride-silicon structures fabricated using these layers. The *charge distribution* in the layers provides important information on the origin of these charges because it can be related to the distribution of various bonds as observed by infra-red absorption measurements.

In this paragraph measurements will be reported on layers deposited in a single step and on layers deposited in several subsequent steps. The experiments on multi-step deposited wafers were conducted in order to obtain more information on the effect of the start of the deposition process and of the role of the Si-substrate on the characteristics of the layers. The total charge in the layers and the charge distributions were studied both on as-deposited layers and after several anneal treatments.

A. Measurement of the charge distribution

The charge distribution in an insulating layer can be determined from the measurement of the flatband voltage (V_{fb}) as a function of the thickness of the layer (x_n) [1]. In order to construct this diagram the layers are step-wise etched and capacitor structures are fabricated by the deposition of aluminium and standard wet lithography. In order to increase the number of measurement points the step-wise etching is performed in two directions on the wafer which are perpendicular to each other. This generates a pattern on the wafer as indicated in fig.1. The thickness of the layers decreases (or increases) from left to right and from top to bottom of the wafer.

The charge distribution in the layers can be calculated through the general expression of the flatband voltage [1] :

$$V_{fb} = \Phi_{MS} - t_{oxn}.Q_{it}/\varepsilon_{oxn} - (1/\varepsilon_{oxn}).\int_0^{t_{oxn}} x.\rho(x).dx \qquad (1)$$

with Φ_{MS} the workfunction difference between the gate metal and the silicon substrate

t_{oxn} the thickness of the oxynitride layer

ε_{oxn} the dielectric permittivity of the oxynitride layer

Q_{it} the charge in the interface traps at a surface potential zero (corresponding to V_{fb})

x the location in the oxynitride layer measured from the gate electrode

$\rho(x)$ the charge distribution in the oxynitride layer.

The first derivative of the V_{fb}-vs-x_n curve provides information on the total product of the charge in the layer and its distance from the Si/oxynitride interface. A negative slope of this curve indicates that the dominant charge between the Si/oxynitride interface and x_n is positive, while a positive slope indicates that negative charge is dominant in this region. The second derivative of the V_{fb}-vs-x_n curve provides information on the local charge in the layer. When this derivative is zero no local charge is present. A positive value indicates local negative charge and a negative derivative indicates local positive charge.

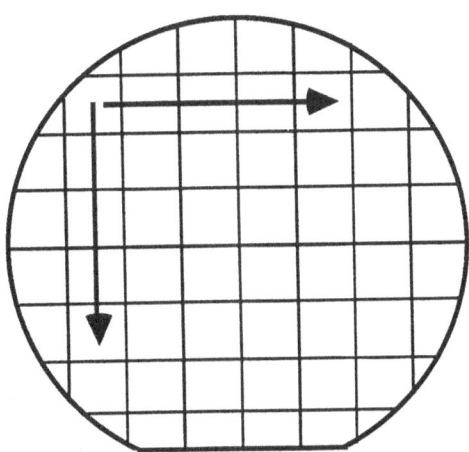

Fig. 1 : *Pattern on the wafer generated by the step-wise etching procedure in two directions. The thickness of the layer decreases (or increases) from left to right and from top to bottom of the wafer. The arrows indicate the approximate centroid of the negative compensating charge.*

B. One-step deposited layers

a. As-deposited layers

Oxynitride layers were deposited with various O/(O+N) ratios to a thickness of approximately 100 nm. The results of the charge profiling measurement on the as-deposited layers is shown in fig.2. The negative slope of the V_{fb}-vs-x_n, found for all O/(O+N) ratios at layer thicknesses smaller than 25 nm, indicates that in all layers a large positive charge (Q^+) is present near the Si/oxynitride interface. From the similarity of these initial slopes for all O/(O+N) ratios it can be tentatively concluded that this positive charge is identical for all measured samples.

As the second derivative is zero in the bulk of the layer for all O/(O+N) ratios smaller than approximately 0.35 (the V_{fb}-vs-x_n curves can be approximated by a straight line) the bulk of the oxynitride is free of charge for these layer compositions.

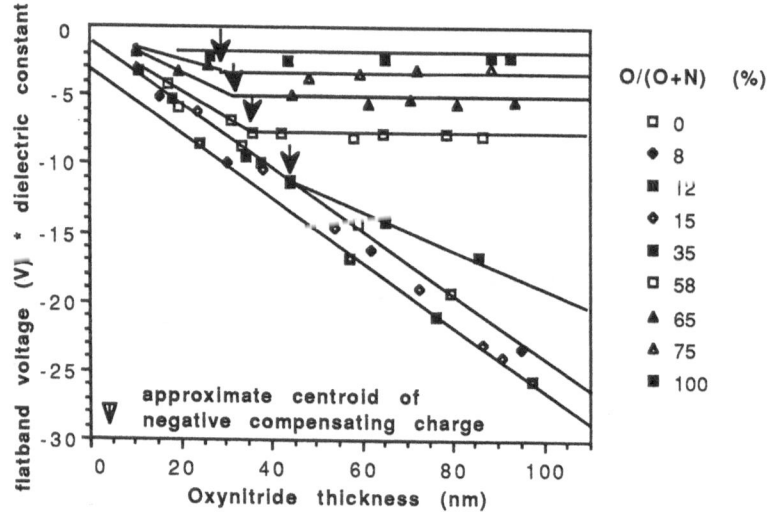

Fig. 2 : *Flatband voltage-vs-thickness measurements for as-deposited oxynitride layers with various O/(O+N) ratios.*

For an O/(O+N) ratio of 0.35 a spatially localized negative bulk charge (Q^-) starts to appear which for larger O/(O+N) ratios exactly compensates the positive near-interface charge, indicated by the zero slope of the V_{fb}-vs-x_n curves. The centroid of the spatially localized negative charge has a clear tendency to move to the Si/oxynitride interface with increasing oxygen content of the layer : it changes from approximately 45 nm at an O/(O+N) ratio of 0.35 to 25 nm at an O/(O+N) ratio of 0.75.

b. Reproducibility of the charge distribution

The reproducibility of the charge distribution and its dependence on the thickness of the deposited layer was checked for an O/(O+N) ratio of 0.58. This ratio was chosen because the breakpoint in the V_{fb}-vs-x_n curve (which demonstrates the presence of a spatially localized negative bulk charge) is most prominent for this oxynitride composition. This is clearly evidenced in fig.2. Two extra samples (referred to as samples "B" and "C") with this O/(O+N) ratio and with a thickness of approximately 100 nm were prepared in two separate depositions. The results on the charge distribution in these samples were compared with the result obtained in the first series for this O/(O+N) ratio (referred to as sample "A"), as reported in fig.2.

The resulting V_{fb}-vs-x_n diagrams on samples B and C are shown in fig.3. From these results it can be concluded that on the as-deposited samples the centroid of the localized negative charge is nearly identical for the sample A (fig.2) and samples B and C (fig.3). The absolute magnitude of the negative and the positive charge densities, however, is strongly different in the three samples. While in sample A the localized negative charge exactly compensates the positive near-interface charge, this is not the case in samples B and C. On these samples a smaller density of positive near-interface charge was found and the localized negative charge overcompensates the positive charge, indicated by the positive slope of the V_{fb}-vs-x_n curve for thicknesses larger than 40 nm. The density of this negative charge was also different for the two samples.

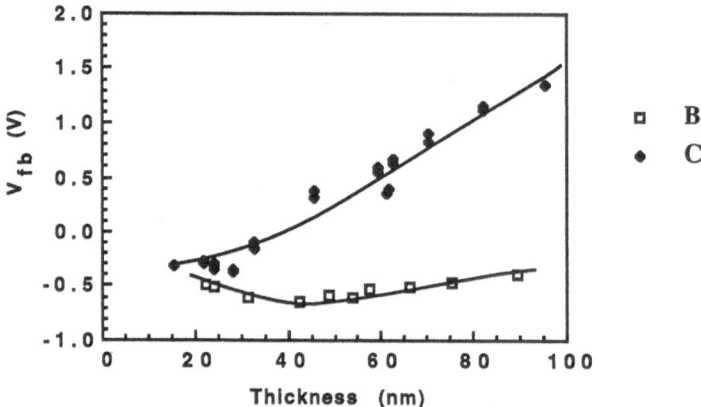

Fig. 3 : *Flatband voltage-vs-thickness measurements for two as-deposited oxynitride layers with an O/(O+N) ratio of 0.58. The oxynitride layers were deposited in two different runs.*

In general this demonstrates that the exact compensation of the positive charge by the localized negative charge, as was found in the as-deposited sample with an O/(O+N) ratio of 0.58 (and found for all O/(O+N) ratios larger than 0.35) of the first series, is not an intrinsic property of the oxynitride layer. Most probably the exact magnitude of the charge in the layers is strongly dependent on small details of the deposition procedure which are not (yet) well controlled and on the interface quality at the start of the deposition.

c. Thickness dependence of the charge distribution

The effect of the starting oxynitride thickness on the charge distribution was checked by depositing three layers with an O/(O+N) ratio of 0.58 and with varying thickness (approximately 50 nm, 100 nm and 150 nm). From the measurements shown in fig. 4 it can be concluded that the position of the breakpoint (where the positive charge is compensated by the localized negative charge) is varying with the initial thickness of the deposited layer. This breakpoint is found to move away from the Si/oxynitride interface with increasing thickness of the layer. This suggests that the density and/or the location of the defects responsible for these charges is changing during the deposition of the layer.

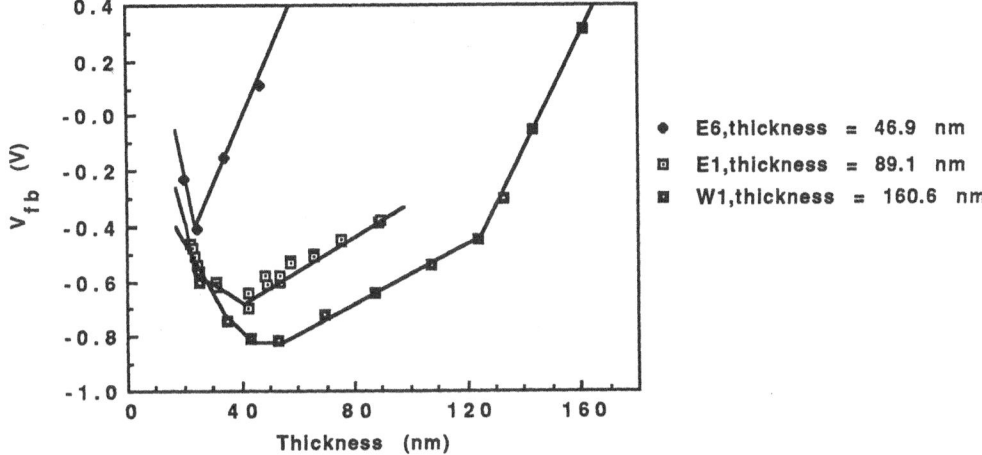

Fig. 4 : *Effect of the initial oxynitride thickness on the flatband voltage-vs-thickness measurements for as-deposited oxynitride layers with an O/(O+N) ratio of 0.58.*

d. Effect of anneal treatments on the total charge in the layers

The total charge density (Q_{ox}) calculated from capacitance-voltage measurements is the sum of the relative contributions (depending on the distance from the Si/oxynitride interface) of Q^+ and Q^-. When measurements are performed on only one layer thickness a straightforward interpretation of the results in terms of charge distributions is, therefore, not possible. It is, however, still interesting to investigate the change in the flatband voltage after various anneal treatments. Therefore, a series of samples were given (successive) post-deposition treatments as described in paragraph 2. The flatband voltage was measured on the as-deposited samples and after each anneal step. The results are summarized in table 1. It must be noted that a negative flatband voltage is an indication for the presence of positive charge (and, accordingly, a positive flatband voltage for negative charge).

It is interesting to note that the initial large positive charge density found on the as-deposited layers becomes negative after an N_2 anneal for layers with an O/(O+N) smaller than 0.34. This anneal was

O/(O+N)	as-dep.	N$_2$ an.	+H$_2$ an.	+Pl. an.	NH$_3$ an.
0.00	-.029	.0105	-.019	-.011	-.022
0.19	-.036	.0094	-.032	-.018	-.025
0.24	-.032	.014	-.040	-.022	-.041
0.34	-.027	.025	-.043	-.043	-.030
0.52	-.026	-.023	-.038	.0024	-.011

Table 1 : *Initial flatband voltages normalized to the layer thickness (V/nm) for different oxynitrides after various successive heat treatments. A negative flatband is an indication for the presence of positive charge (and, accordingly, a positive flatband voltage for negative charge).*

also found to strongly reduce the hydrogen content of the oxynitride layer (chapter 1). A subsequent H$_2$ anneal increases the positive charge and was found to partially restore the hydrogen content in the layer. As the total charge in the layers is dominated by the positive interface charge, these results strongly suggest the existence of a correlation between the positive near-interface charge and the hydrogen content of the oxynitride for O/(O+N) ratios smaller than 0.34. For O/(O+N) ratios larger than 0.34 the total charge in the layer remains positive after an N$_2$. anneal and only slightly decreases. A subsequent H$_2$ anneal further increases the positive charge.

A plasma-H anneal causes a small reduction of the positive charge, while an NH$_3$ anneal has only a minor effect on the total charge. As compared to the non-annealed samples, a small decrease is seen in the positive fixed charge density after the NH$_3$ anneal, except for the wafers with an O/(O+N) ratio of 0.26 and 0.33, where an increase in the fixed charge density is seen.

e. Effect of anneal treatments on the charge distribution

Because it would be an enormous task to construct a diagram of V_{fb}-vs-x_n after all applied anneal treatments (as was done for nitrided layers in ref. 1) and for all O/(O+N) ratios in order to measure the charge distribution in the oxynitride layers, this experiment was only performed on layers with an O/(O+N) ratio of 0.58. This composition was used because of the clear presence of localized negative charge in these layers, evidenced in fig. 2. The resulting Vfb-vs-xn diagrams after the various anneal treatments (described in paragraph 2) are shown in fig.5. A schematic of the charge distribution in the layer, as inferred from these results, is shown in fig.6.

After an N_2 anneal the positive charge of the as-deposited sample still exists near the interface (fig.6a and 6b). The localized negative charge, however is no longer present after this anneal. Two possible explanations for this effect are : i) the negative charge is closely related to the presence of hydrogen (and the prime effect of the N_2 anneal is the removal of hydrogen) or ii) the higher temperature of the anneal step causes a local structural re-arrangement whereby the defect responsible for the negative charge is annealed. In this respect it must be noted that the temperature during the anneal treatment (1000°C) is larger than during the deposition (800°C).

A subsequent H_2 anneal introduces positive charge with a decreasing density from the outer surface towards the oxynitride bulk (fig.6c). This observation can be viewed as an extra indication for the relation between positive charge and the presence of hydrogen (most likely in Si-H groups) as this profile can be expected to result from the inward diffusion of hydrogen.

After an NH_3 anneal negative charge is found to dominate in the oxynitride layer (fig.6d). However, as the total flatband voltage was found to be negative after this anneal (table 1), a large density of positive charge must be present near the Si/oxynitride interface. The results, however, suggest that nitrogen bonded to hydrogen is not directly responsible for positive charge in the oxynitride layers.

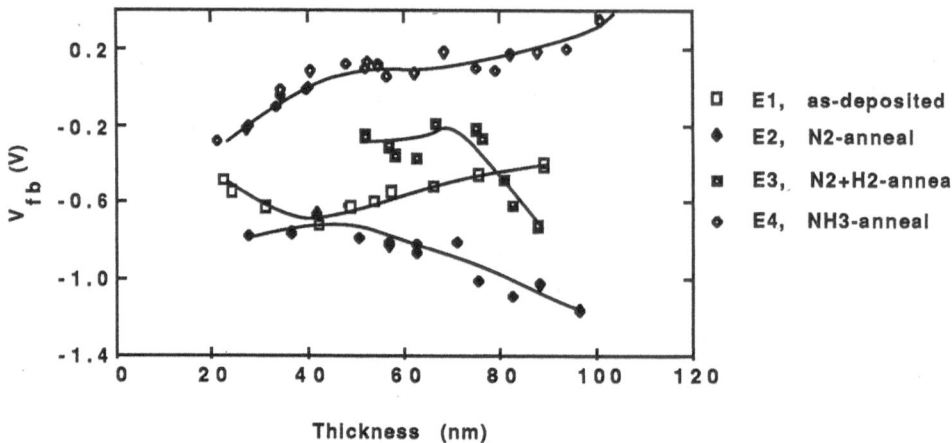

Fig. 5 : *Flatband voltage-vs-thickness measurements for a 100 nm thick oxynitride layer with an O/(O+N) ratio of 0.58. Results are shown for the as-deposited layer, after an N2 anneal, after an N2+H2 anneal and after an NH3 anneal.*

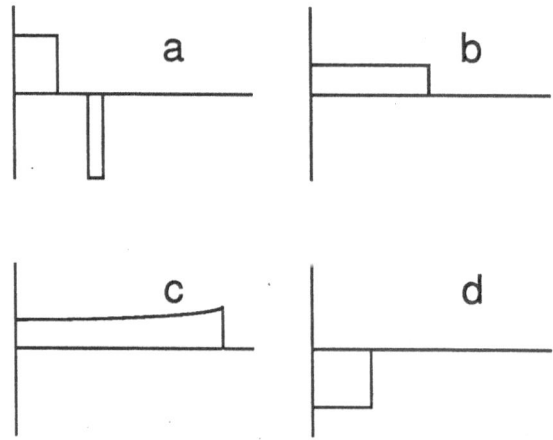

Fig. 6 : *Schematic representation of the charge distributions in the one-step deposited oxynitride layers with an O/(O+N) ratio of 0.58 as calculated from the measurements shown in fig.5. Results are shown for the as-deposited layer (a), after an N2 anneal (b), after an N2+H2 anneal (c) and after an NH3 anneal (d).*

C. Multi-step deposited layers

a. As-deposited layer

In order to determine whether the pile-up of positive charge near the Si/oxynitride interface is due to the effect of the silicon substrate on the deposited film or to the occurrence of a non-steady state phenomenon during the initial stage of deposition, a 150 nm thick oxynitride layer with an O/(O+N) ratio of 0.58 was deposited in three sequential steps (without removal of the sample from the LPCVD tube) of approximately 50 nm each. The wafer was taken through a sequence of etch-back steps and V_{fb}-measurements in order to determine the charge distribution in the oxynitride layer.

If the positive near-interface charge and the localized negative charge are generated by a non-steady state phenomenon occurring in the initial stage of the deposition it can be expected that this charge distribution is also found at the boundary of the sequentially deposited layers. If, however, the charge distribution results from the effect of the Si-substrate (and/or the native oxide on the Si-surface) on the oxynitride deposition process, charge-free boundaries between the layers should be formed. The V_{fb}-vs-thickness relation measured on the wafers deposited in three steps is shown in fig.7.

The curve of the three-step as-deposited wafer exhibits two 'discontinuities' dividing the curve into three regions, which could possibly be interpreted in terms of local positive and negative charge between two subsequently deposited layers. However, both discontinuities coincide with the change of position from one side of the wafer to the other (fig.1) and, because on as-deposited layers non-uniformities in the flatband voltage in the order of 0.2 V over a 4 inch wafer were found, these discontinuities are most likely to be an artifact. It can, therefore, be suggested that near the interface of two subsequently deposited layers no (measurable) charge is present in the oxynitride layers.

The positive near-interface charge which is always observed (also in these multi-step deposited layers) is, therefore, directly related to the presence of the Si-substrate. This means that either the crystalline structure of the Si-substrate imposes a certain structure (and defect type) in the near-interface region of the oxynitride layer (at least for this O/(O+N) ratio) or the native oxide present on the Si-surface plays an important role at the start of the deposition. This is an important observation which will be used in chapter 5 to propose a model for the charge distribution in these layers.

b. Effect of anneal on the charge distribution

The charge distribution after various anneal treatments (equivalent to those used in paragraph 2) was also investigated on these multi-step deposited layers. The results are also shown in fig.7.

After an N_2 anneal approximately the same amount of positive charge near the interface is found as on the as-deposited wafer, as evidenced by the equal slope of the V_{fb}-vs-x_n curve for both samples. Consistent with the observations for the samples with a one-step deposition the localized negative charge is (mostly) removed during the N_2 anneal. An H_2 anneal after the N_2 anneal causes a decrease in the positive charge near the interface. No information could be obtained near the outer surface due to problems during the etching which meant

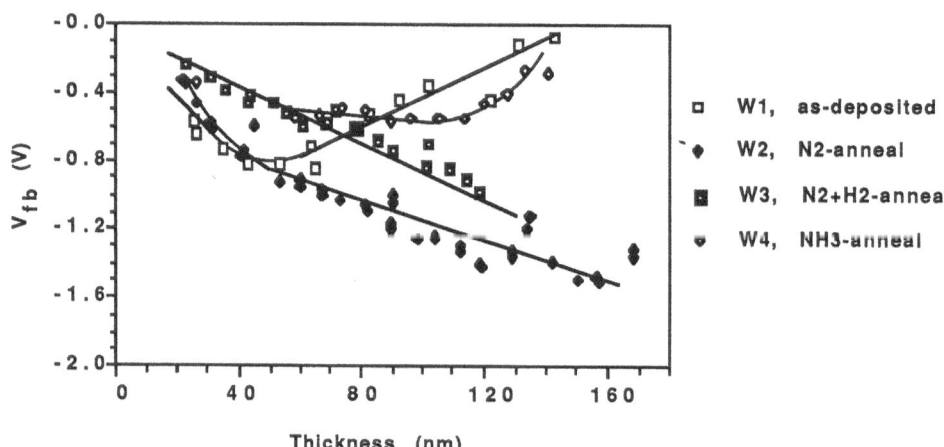

Fig. 7 : V_{FB}-vs-thickness measurements for the three-step deposited oxynitride layer with an O/(O+N) ratio of 0.58. The layer was deposited in three sequential steps of approximately 50 nm each. Results are shown for the as-deposited layer, after an N_2 anneal, after an N_2+H_2 anneal and after an NH_3 anneal.

that only 2/3 of the layer thickness could be measured. These results are consistent with the results found on the one-step deposited layer (cfr. fig.5 and 6).

After an NH_3 anneal the near-interface charge is found to be positive with a magnitude which is smaller than that found in the as-deposited layer. The positive charge can be seen from the negative slope of the V_{fb}-vs-x_n curve. This result is different from the result obtained on the one-step deposited layer (in fig.5 a negative charge was found after this anneal). This can be related to the different thickness of the layer which affects the diffusion of NH_x towards the Si/oxynitride interface.

4. INTERFACE TRAP DENSITY

A. Interface trap density at midgap

The interface trap density as a function of the Si-bandgap was determined from the quasistatic and high-frequency capacitance-vs-voltage (CV) measurements [2]. Measurements were performed on both as-deposited and annealed samples.

a. As-deposited layers

The interface trap density at midgap (D_{it}) on the as-deposited layers as a function of the O/(O+N) ratio shows a rather narrow transition region around an O/(O+N) ratio of 0.4 where the interface trap density suddenly drops from the 10^{12} traps.$cm^{-2}eV^{-1}$ range to values of a few times 10^{10} traps.$cm^{-2}eV^{-1}$. This result is shown in fig. 8. Obviously, very low interface trap densities can be obtained by adding sufficient amounts of oxygen to the film.

The sudden change in the Si/oxynitride interface quality at an O/(O+N) ratio of 0.35 is an indication for either a change in the layer structure at this composition or a sudden change in the hydrogen concentration at the Si/oxynitride interface. A different layer structure

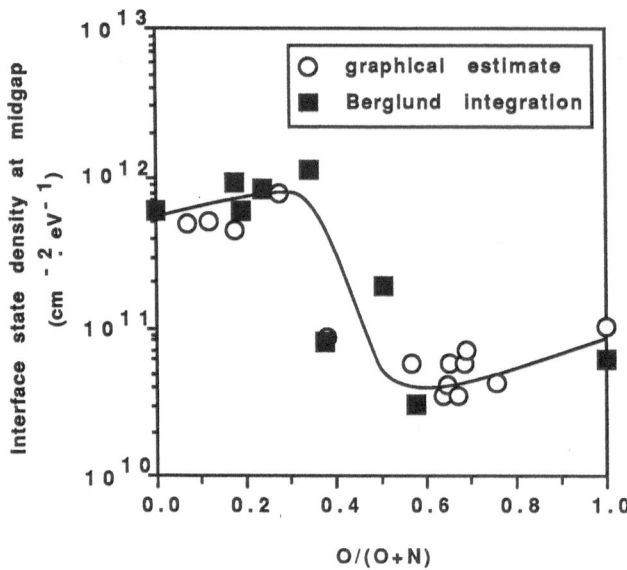

Fig. 8 : *Interface trap density at midgap (as calculated from the quasistatic CV-curves) as a function of the O/(O+N) ratio of the oxynitride layer. Results for two separate runs are shown as indicated by the different symbols.*

at an O/(O+N) ratio of 0.35 (or larger) could result in a better matching between the Si-substrate and the oxynitride film, giving rise to a lower density of dangling bonds and, therefore, a lower interface trap density. A larger hydrogen concentration at the Si/oxynitride interface for these O/(O+N) ratios could explain the lower interface trap density by an effective neutralization of the dangling bonds by the hydrogen, rendering the interface defect inactive as an interface trap [3-5].

b. Effect of anneal treatments

The midgap interface trap density after various anneal treatments (as described in paragraph 2) is shown in fig.9. An N_2 anneal increases the interface trap density at midgap for O/(O+N) ratios smaller than 0.35 while for larger O/(O+N) ratios a decrease is found. At the transition point (O/(O+N)=0.35) the interface trap density at midgap is unaltered by such an anneal. These results are, again, an indication of a structural change in the Si/oxynitride interface occurring at an O/(O+N)

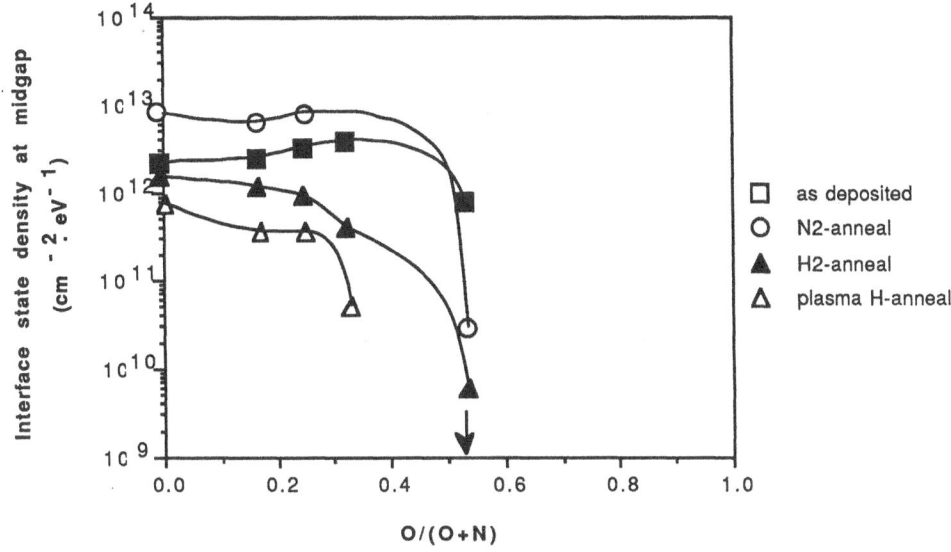

Fig. 9 : *Interface trap density at midgap (as calculated from the quasistatic CV-curves) as a function of the O/(O+N) ratio of the oxynitride layer and of the post-deposition anneal. Results are shown for the as-deposited layers, after an N_2 anneal, an N_2+H_2 anneal and a subsequent plasma-H anneal.*

ratio of 0.35. A subsequent H_2 anneal and an H_2 plasma anneal both lower the midgap interface trap density for all O/(O+N) ratios. This can be ascribed to the effect of hydrogen which passivates the interface traps.

In a seperate experiment it was found that an NH_3 anneal results in a relatively strong decrease of the interface trap density. The final interface trap density is comparable with the density found after an H_2 anneal, indicating the important role of hydrogen during the NH_3 anneal.

B. Distribution of the interface trap density

The distribution of the interface trap density as a function of the silicon bandgap shows a maximum above midgap for O/(O+N) ratios

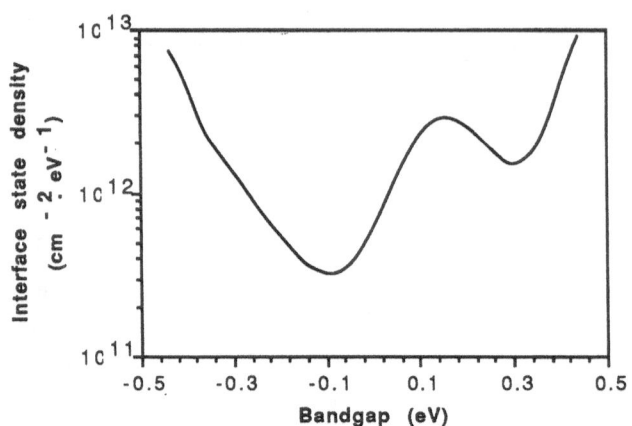

Fig. 10 : *Interface trap density as a function of the Si-bandgap for an oxynitride layer with an O/(O+N) ratio of 0.33 after an NH₃ anneal. A peak in the interface trap density is observed at 0.17 eV above the middle of the Si-bandgap.*

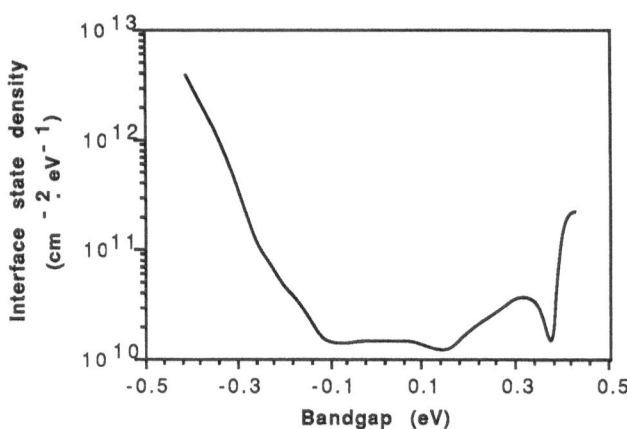

Fig. 11 : *Interface trap density as a function of the Si-bandgap for an oxynitride layer with an O/(O+N) ratio of 0.52 after an NH₃ anneal.*

smaller than or equal to 0.35. An example of a typical distribution with a peak is given in fig. 10 (O/(O+N)=0.33), while fig. 11 gives a typical exemple of a curve without the peak (O/(O+N)=0.5). Both results were obtained after an NH₃ anneal. The height of the interface trap density peak varies in the same manner as the midgap interface trap density during the different anneal treatments. This is demonstrated in fig. 12 where the effect of anneal treatments on the interface trap distribution is shown for a specific (but arbitrarily choosen) oxynitride. The energy position of the peak depends on both the O/(O+N) ratio and the anneal treatment but is typically located around 0.17 eV above the middle of the Si-bandgap.

The peak in the interface trap density is commonly related to silicon dangling bonds at the interface [6-9]. The shift of the peak position with varying O/(O+N) ratio can be ascribed to the different environment of the dangling bond with changing composition and/or structure of the layer and is, therefore, dependent on both the O/(O+N) ratio of the layer and the local hydrogen concentration.

C. Correlation between D_{it} and Q_{ox}

It is important to notice that the O/(O+N) ratio of 0.35 is a transition point for the characteristics of both D_{it} and Q_{ox} and for the interrelation between these properties. This demonstrates the basic structural change occurring at this O/(O+N) ratio.

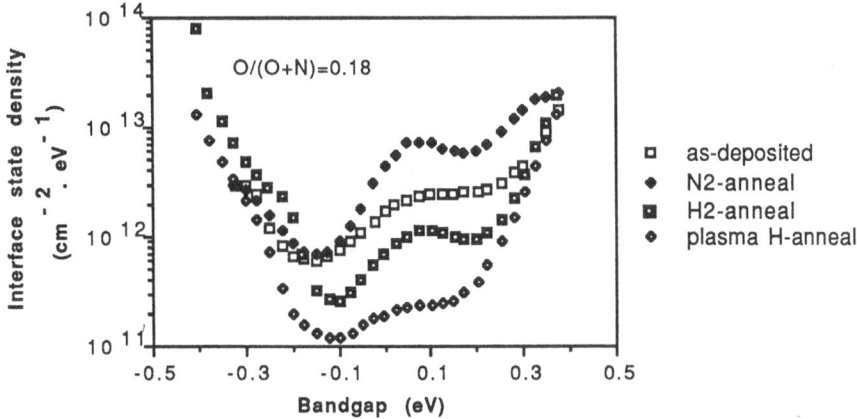

Fig. 12 : *Interface trap density as a function of the Si-bandgap for an oxynitride layer with an O/(O+N) ratio of 0.18 after various anneal treatments.*

For O/(O+N) ratios smaller than 0.35 the variations in D_{it} and in Q_{ox} upon different anneal treatments show remarkable similarities which can be correlated with the varying hydrogen content of the oxynitride films (Chapter 1). A decrease of the hydrogen content during an N_2 anneal leads to an increase of D_{it} (fig.9) and a change from positive to negative Q_{ox} (table 1). This is tentatively ascribed to a decrease of Q^+ related to Si-H bonds. A subsequent H_2 anneal will increase the hydrogen concentration in the film and will generate the opposite effects.

The interface trap density for O/(O+N) ratios larger than 0.35 is much smaller and seems to be independent of the hydrogen concentration for these O/(O+N) ratios because each anneal treatment results in a decrease in the measured D_{it}-values (fig.9). Therefore, no direct correlation between the D_{it} and Q_{ox} values can be observed for O/(O+N) ratios larger than 0.35.

D. Effect of the oxynitride deposition on the interface trap density of an underlying SiO2 layer

Oxynitride layers were deposited on thermally grown oxide layers with a thickness of respectively 3 nm and 30 nm. The processing details are given in section 2 of this chapter.

A 30 nm oxide between the Si-substrate and the oxynitride layer is found to act as an efficient barrier to protect the Si/SiO₂ interface quality during deposition and the interface density on all samples with a 30 nm oxide layer (all O/(O+N) ratios and all anneal treatments) is in the 10^{10} traps/cm² range, typical for thermal oxides. Only minor differences are found between the various samples which are difficult to reliably resolve within the measurement accuracy of the quasistatic CV-method. Only for the NH_3 annealed samples a small increase in the interface trap density can be evidenced, which is eventually related to a small nitridation of the oxide layer.

The presence of a 3 nm oxide layer changes the interface trap density to values between the typical values for a thermal oxide and the values found for the oxynitride/Si-interface. When it is assumed that the 3 nm oxide layer is nitrided during the first stages of the deposition [10] this behaviour can be expected. On these samples both the N_2 and the H_2 anneal cause a decrease in the interface trap density for all O/(O+N) ratios. As this behaviour was also observed for the interface between the Si-substrate and oxynitride layers with an O/(O+N) ratio larger than 0.35, this suggests that the O/(O+N) ratio of the nitrided 3 nm oxide layer is larger than 0.35. A significant increase in the interface trap density is found after an NH_3 anneal for oxynitride layers with an O/(O+N) ratio smaller than 0.35. This is probably due to a further nitridation of the oxide layer under these conditions.

5. BULK PROPERTIES

A. Relative permittivity

The relative permittivity of the layers was calculated from high-frequency CV-curves. The measurements were performed at either 10 kHz or 100 kHz in order to avoid errors due to resistance effects. The relative permittivity of the layers as a function of the O/(O+N) ratio is given in fig. 13. A smooth variation from 7.0 (Si_3N_4) to 3.8 (SiO_2) was found.

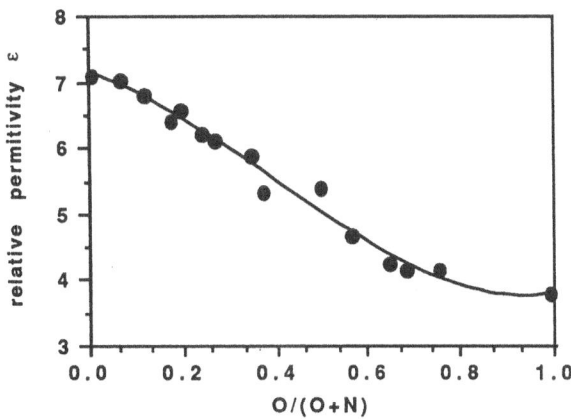

Fig. 13 : *Relative permittivity of the oxynitrides as a function of the oxygen content of the layers.*

B. Charge trapping

From the measured saturation values of the flatband voltage (V_{fb}) shifts upon the application of a constant injection field across the tunnel oxide the trapped charge density in the oxynitride layer in stacked oxide/LPCVD oxynitride structures is found to decrease with increasing oxygen content of the layers. This is thought to be an indication of a decreasing trap density with increasing O/(O+N) ratio. Because the maximum V_{fb} shift is constant over the different O/(O+N) ratios up to a value of 0.35 the trapped charge density is proportional to the dielectric constant of the oxynitride. For larger O/(O+N) ratios the trapped charge density decreases faster than the proportionality with the dielectric constant would predict.

C. Conduction

a. I-V measurements

Current-vs-voltages curves reveal that, as the oxygen content increases, the conductivity of the layers decreases (fig.14), the largest

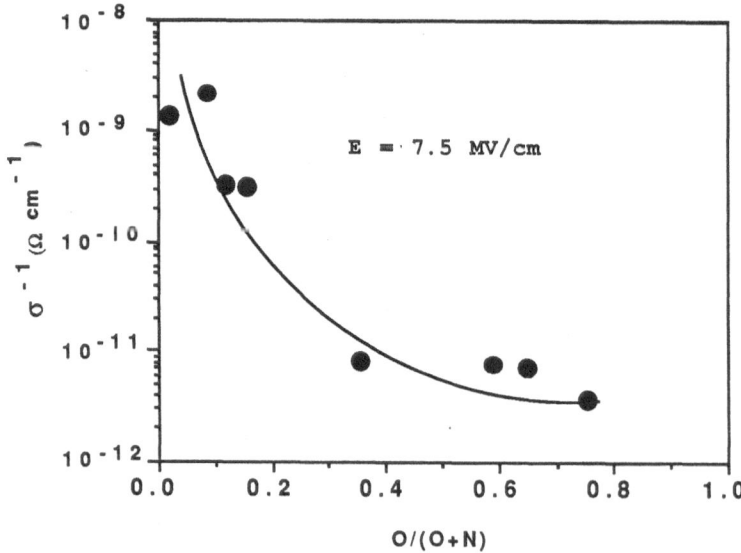

Fig. 14 : *Conductivity of the oxynitrides as a function of their composition for a fixed mean field of 7.5 MV/cm.*

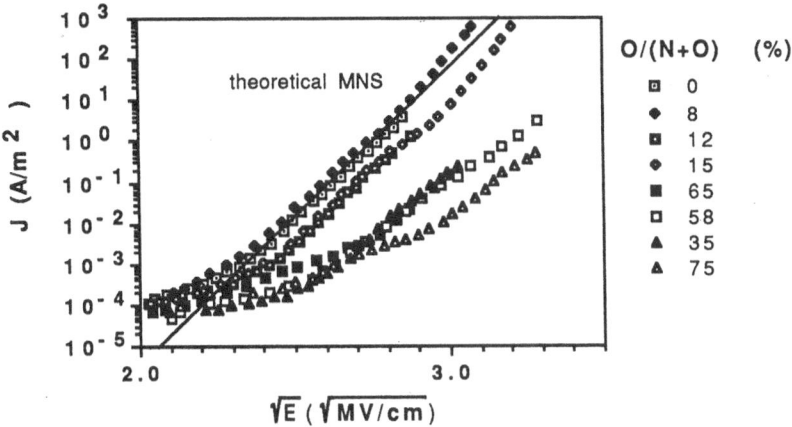

Fig. 15 : *Current-voltage characteristics of oxynitride layers with various compositions plotted in a Poole-Frenkel plot.*

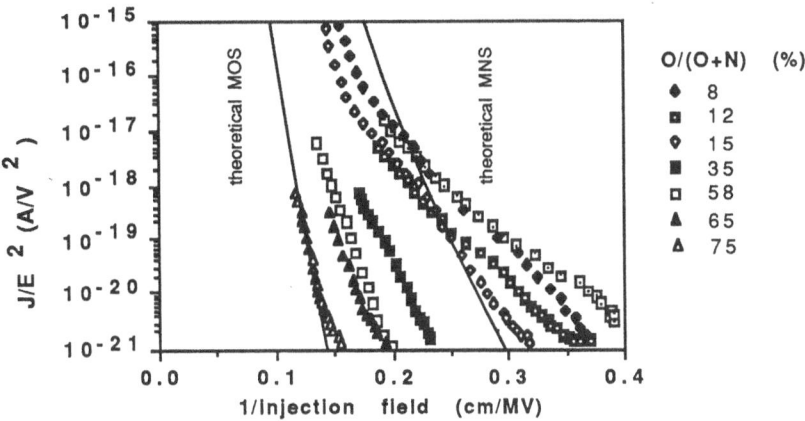

Fig. 16 : *Current-voltage characteristics of oxynitride layers with various compositions plotted in a Fowler-Nordheim plot.*

variation (again) occurring at about O/(O+N)=0.35 [11]. The fit of the current-voltage curves to a Poole-Frenkel plot improves upon increasing the nitrogen content of the layer (fig.15). For specimens containing more nitrogen an increasing deviation from the Fowler-Nordheim curve is observed (fig.16).

The effect of annealing on the current-vs-voltage measurements is very restricted. An N_2+H_2 anneal decreases (increases) the current of the samples with O/(O+N) smaller (larger) than 0.34 and does not change the current of the sample with O/(O+N)=0.34 compared to the current on the as-deposited samples (fig.17). A plasma H_2 anneal reduces the conductivity. This may indicate the removal of shallow conduction assisting traps from the bandgap or, presumably, the effect of a reduced injection efficiency due to the considerable D_{it}-lowering.

a)

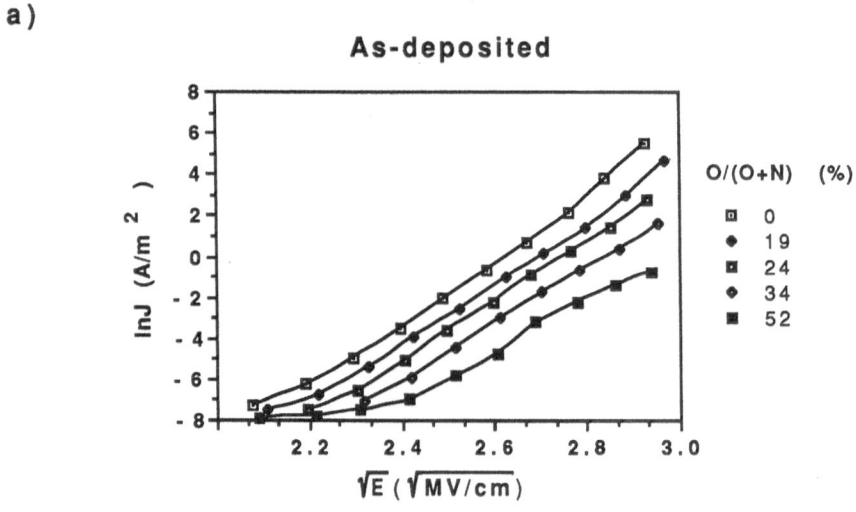

b)

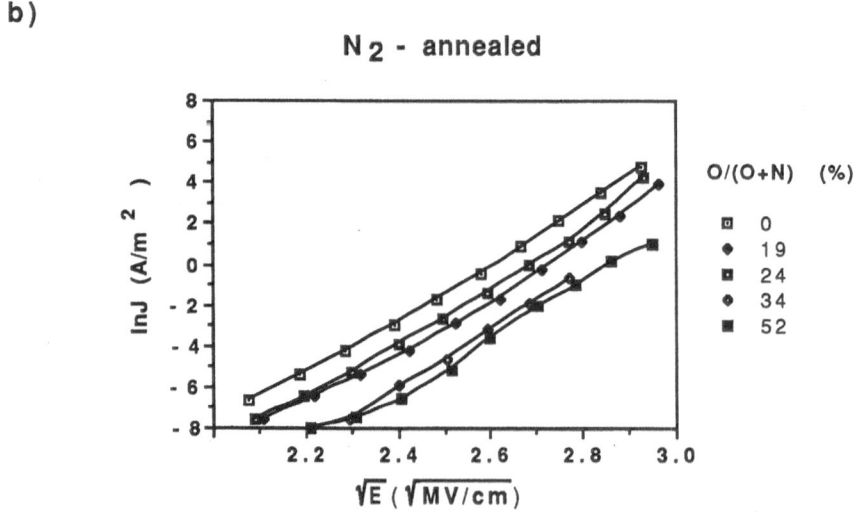

c)

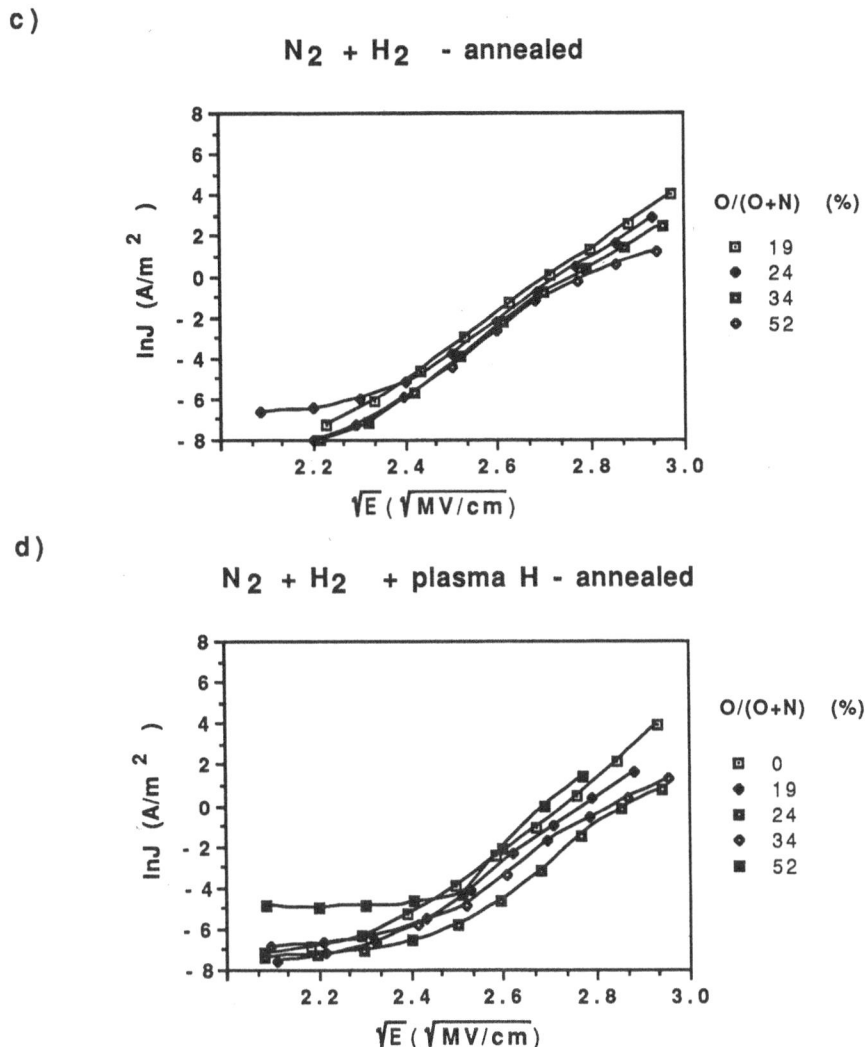

N₂ + H₂ - annealed

N₂ + H₂ + plasma H - annealed

Fig. 17 : *Current-voltage curves of oxynitrides with various composition after several anneal sequences: a) as-deposited; b) after N₂-anneal; c) after N₂+H₂-anneal and d) after N₂+H₂+plasma H anneal.*

b. Weinberg measurements

The conduction through the oxynitride layers was studied using Weinberg measurements [12]. To this purpose specially designed test structures were fabricated [13]. A cross-section of the device is shown in fig. 18. The gate insulator, consisting of the material under investigation, is found in the center of the device and is referred to as the emitter. Underneath the gate a shallow layer (<0.3μm) of the opposite type of the substrate is obtained by ion implantation. This layer is contacted by a ring-shaped diffusion, referred to as the base. The substrate itself is contacted by two half-ring-shaped diffusions, called collector. Five oxynitrides of different compositions were used. The O/(O+N) ratio were 0, 0.14, 0.2, 0.35 and 0.43. The thickness of the oxynitride layers was 35 nm.

On these structures the emitter voltage was stepped in increments of 1V starting from a low positive voltage. At each voltage and after a settling time of 1 minute the three currents I_e, I_c and I_b are measured. During the complete cycle the collector is grounded. After the currents are obtained, the flatband voltage is also measured. Fig. 19 shows a typical result for the layer with an O/(O+N) ratio of 0.35. As can be seen I_b is almost equal to I_e while the collector current is one to two orders of magnitude smaller than the former two currents. This

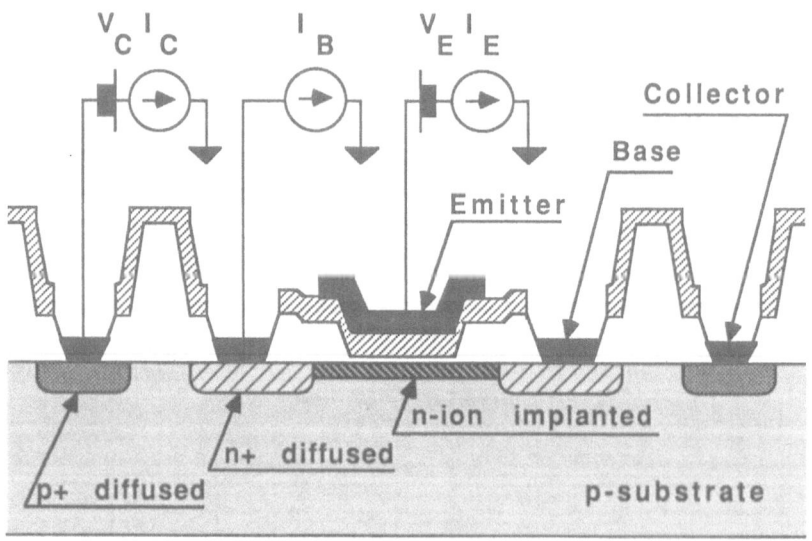

N-channel

Fig. 18 : *Cross section of the devices used to measure the conduction through the oxynitride layers in this study.*

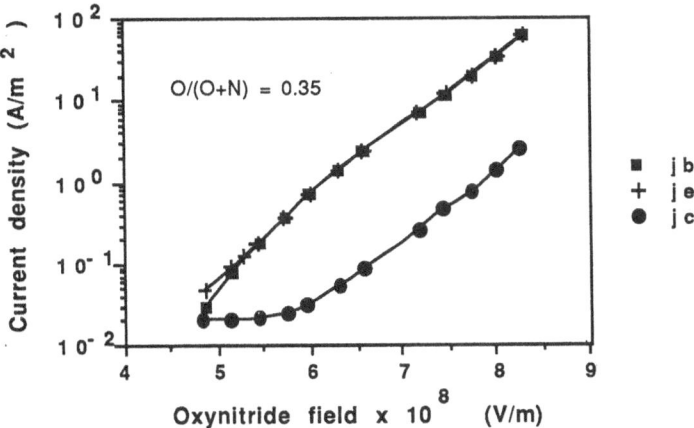

Fig. 19 : *Measured currents as a function of the field in the oxynitride layer with an O/(O+N) ration of 0.35.*

collector current is, however, a clear indication that holes are emitted from the gate structure into the substrate or, in other words, that hole conduction is important in these layers under positive polarities [14-17].

The ratio $\alpha = I_c/I_e$ was found to decrease with increasing electric field. At large fields ($\approx$ 6 MV/cm) its value was found to saturate, which means that the current characteristics of I_c and I_e are parallel. The saturation value for α as a function of the O/(O+N) ratio of the oxynitride layer is shown in fig. 20. It is clear that α increases for increasing oxygen content of the film. The current density for the same electric field in the oxynitride layer is found to decrease with increasing oxygen content in the layer.

From these results some important conclusions can be drawn. It is believed that all results point to dominant hole conduction in the oxynitride layers (at least up to the highest O/(O+N) ratio used in this experiment which is equal to 0.43). The decrease of the current density with increasing oxygen content of the film is due to the higher injection barriers for both electrons and holes. This is consistent with measurements of the photocurrent as a function of the photon energy (for a constant photon flux) where it was demonstrated that the barrier height for electron injection from both the Al-gate and the Si-substrate into the oxynitride decreases continuously with increasing nitrogen content of the oxynitride layer. The same result is obtained from the programming experiments on SO_XNOS transistors to be discussed in Chapter 5.

From the work performed on conduction in silicon nitrides it could be concluded that holes injected from the gate recombine with electrons injected from the silicon close to the Si/SiO_2 interface [15,16]. Although the number of injected holes decreases with increasing oxygen content it is also expected that the recombination of these holes with electrons will decrease for higher oxygen content, hence leading to larger α-values, as is experimentally confirmed in fig. 20. The increase of α for O/(O+N) ratios larger than 0.15 can, therefore, be seen as an indication for a decrease of the trap density with increasing oxygen content of the layer.

D. Flatband voltage-vs-gate voltage

Flatband voltage-vs-injection field curves (V_{fb}-vs-(V_g-V_{fb})) are constructed by applying a gate voltage and, after a certain saturation

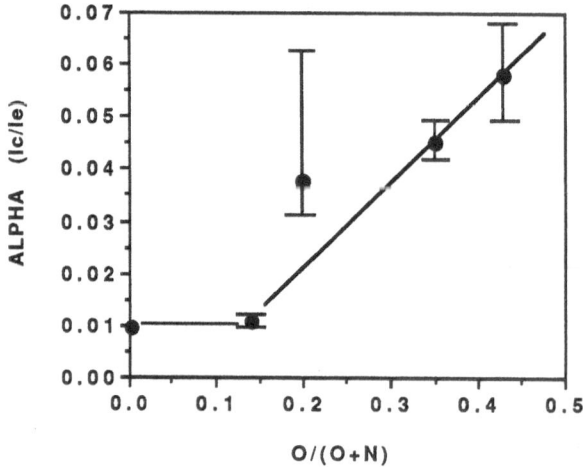

Fig. 20 : *Ratio between the collector current and the emitter current (I_c/I_e) in saturation as a function of the O/(O+N) ratio of the oxynitride layers.*

time, measure the flatband voltage. These curves give an idea of the total charge which can be stored in the film for a certain applied field. The V_{fb}-vs-(V_g-V_{fb}) curves as a function of composition (shown in fig.21) for a fixed film thickness (about 100 nm) indicate the effect of the increasing bandgap width and presumably of the decreasing interface trap density [11]. With increasing oxygen content the substrate contact field required for a significant charge injection (and V_{fb}-shift) increases. With respect to the effect of the interface traps, we draw the attention to the sudden variation of this onset field at about O/(O+N)=0.35, corresponding to the composition at which the interface trap density is strongly reduced.

6. RETENTION

For the use of these layers in MNOS non-volatile memories the decay rate is of extreme importance. Although decay rates were

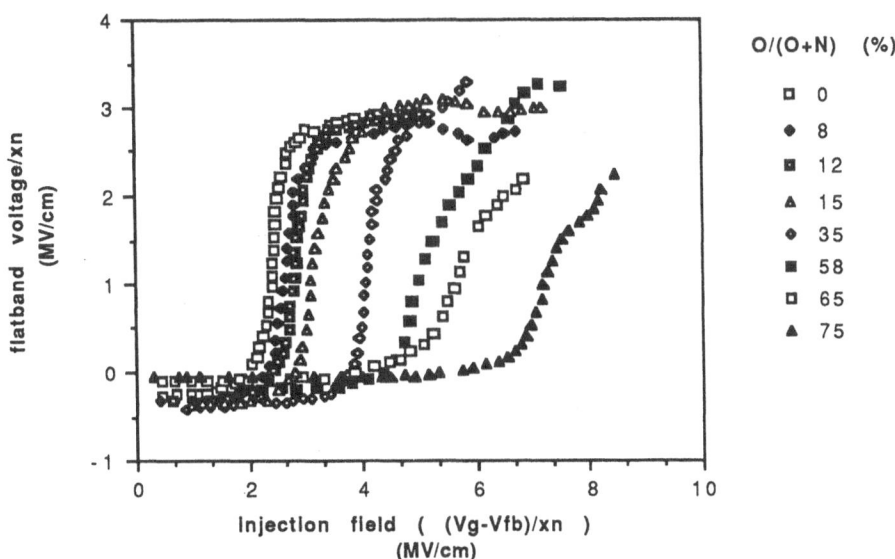

Fig. 21 : *Flatband voltage (V$_{fb}$) versus injection field (V$_g$-V$_{fb}$) curves for oxynitrides with various compositions and a fixed film thickness of 100 nm.*

measured for native underlying oxides only (MNS) -which is not relevant for MNOS operational circumstances- and, because of the lack of annealing in these capacitor structures (in contrast to MNOS transistors, see Chapter 5) are primarily determined by the interface trap densities, the results are mentioned briefly.

For a fixed amount of trapped charge (hence fixed Vfb.n/xn value) the variation of the relative decay rates (dVfb(t)/dlog(t))/Vfb(0) where Vfb(0) is the flatband voltage shift just after removing the gate voltage) with composition is indicated in fig.22. For reasons of accuracy mean values were used obtained on thicknesses >50 nm. After a slight worsening of the relative decay rate with small amounts of oxygen an improvement is found for O/(O+N)>0.3. The similarity between the variations of D_{it} and retention with oxynitride composition is striking and shows that in the present case of capacitor structures, the discharge is primarily determined by the availability of interface traps (fig.23).

For the use in memory transistors one is not merely concerned with a specific charge storage but with a minimum flatband voltage

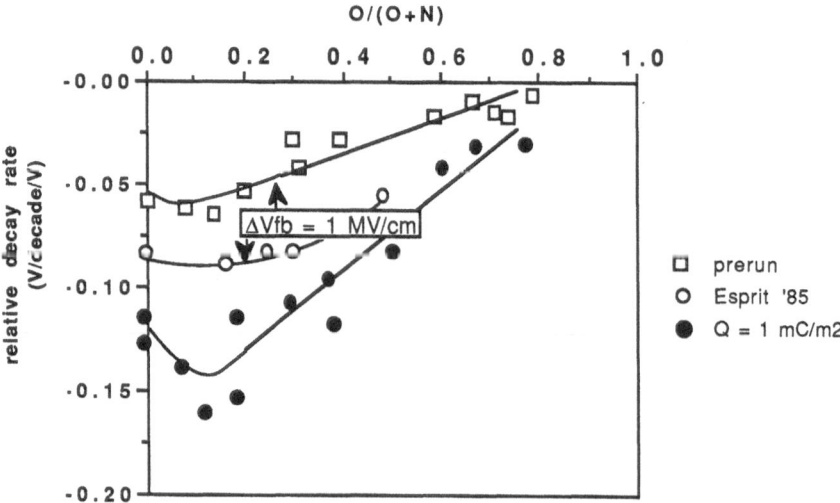

Fig. 22 : *Relative decay rate as a function of the composition for 1 mC/m² electron storage and for 1 MV/cm flatband voltage/thickness shift. The latter values are lower due to a saturation of the decay-ΔVfb relationship at higher ΔVfb.*

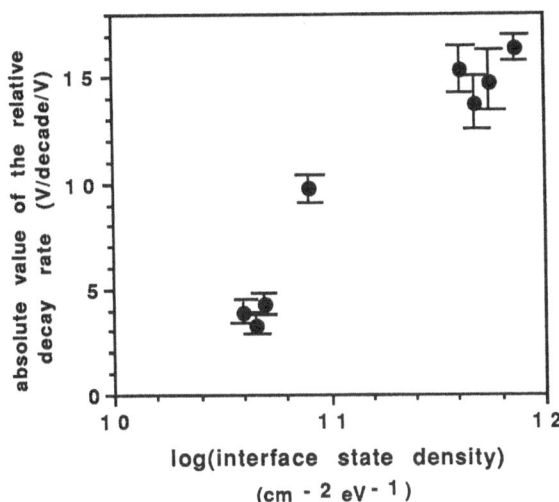

Fig. 23 : *Dominant effect of the interface trap density on the discharge rate.*

shift. Therefore the data of fig.22 also include results for a fixed 1MV/cm Vfb/xn -swing. Similar conclusions can be drawn. In general, as the O-content increases, a given Vfb-shift requires higher programming voltages (in spite of the lower amount of required charge storage) as evidenced from Vfb-Vg curves. This can be attributed to the larger gap width (higher injection field) and in the present case also to the lower interface trap density. The decay rate also improves considerably. This phenomenon is clearly linked to a reduced interface trap density.

7. DIELECTRIC INTEGRITY

The dielectric integrity of the LPCVD-layers was evaluated using breakdown tests. LPCVD-oxynitride layers were deposited with refractive indices corresponding to O/(O+N) ratios of 0.26, 0.33, 0.50 and 1.00. The thickness of the layers was checked by ellipsometry and was for all layers about 35 nm. Breakdown measurements were

Tests of the dielectric integrity of the layers revealed an increasing number of low-field breakdowns for increasing oxygen content of the oxynitride layers. The relatively low number of low-field breakdowns in the films with a low O/(O+N) ratio is related to the high trapping rate in these layers.

In chapter 5 the observations made on the electrical characteristics of the layers, reported in this chapter, will be correlated with the measurements on the physical properties, as reported in the other chapters. From this correlation a model will be proposed which relates the various properties to the structural details of the layers.

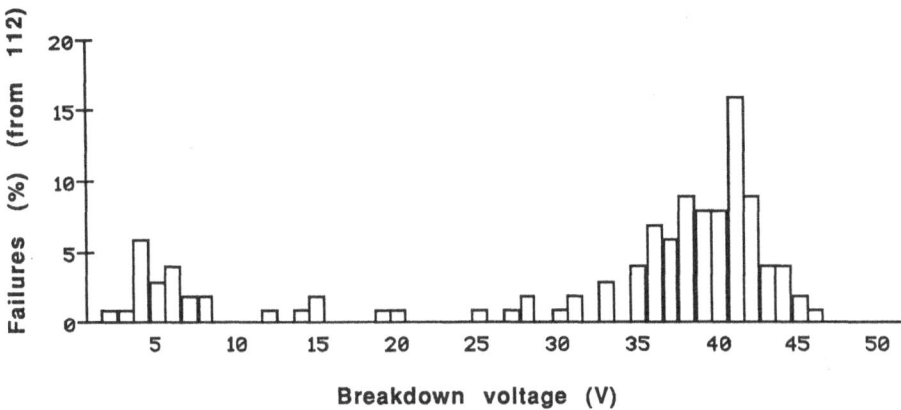

O/(O+N) = 0.33 , as-deposited

Fig. 25 : *Breakdown histogram of an as-deposited oxynitride layer with an O/(O+N) ratio of 0.33 and a thickness of 35 nm.*

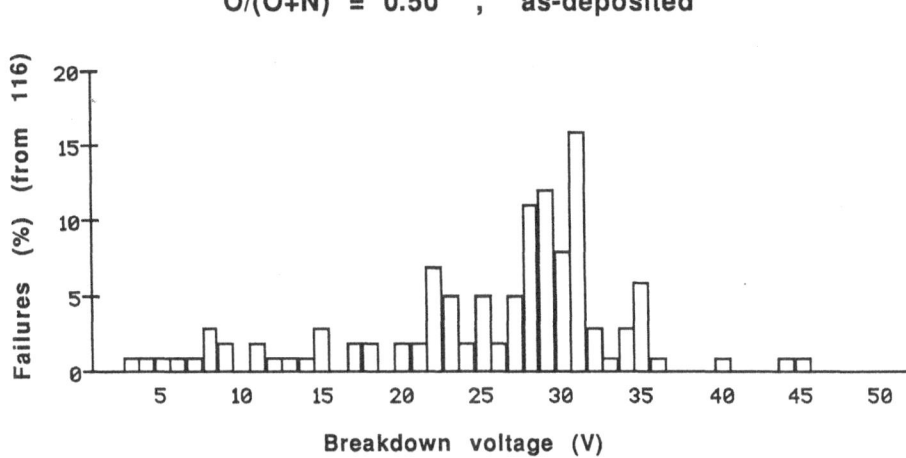

O/(O+N) = 0.50 , as-deposited

Fig. 26 : *Breakdown histogram of an as-deposited oxynitride layer with an O/(O+N) ratio of 0.50 and a thickness of 35 nm.*

detrapping takes place through field detrapping or other wearout and breakdown phenomena start to occur.

The breakdown histogram for the N_2 annealed and N_2+H_2 annealed wafers shows the same trends as the results obtained on the as-deposited samples. Only relatively minor differences in the breakdown histograms were found after the anneal. In detail the effect of the N_2 and N_2+H_2 anneal on the breakdown strength depends complexily on the oxygen content and on the anneal treatment. In general both the N_2 and the N_2+H_2 anneal tend to increase the maximum breakdown strength of the layers.

8. CONCLUSIONS

LPCVD silicon-oxynitrides with compositions ranging from pure oxide to nitride were examined on their electrical properties. The in-depth charge distribution of the layers, the interface trap density, the I-V characteristics and the retention reveal a transition at an O/(O+N) ratio of approximately 0.3-0.4. Other characteristics, such as the breakdown strength, the barrier heigth and the conduction through the layer show a more continuous behaviour as a function of the O/(O+N) ratio.

For all O/(O+N) ratios an almost identical positive near-interface charge is present. For O/(O+N) ratios larger than 0.35 spatially-localized negative charge is present with a centroid which tends to move to the Si/oxynitride interface for larger oxygen contents. It is found that for an identical layer thickness the position of the negative charge centroid is reproducible but not its magnitude. With increasing the layer thickness, the distance between the centroid and the Si/oxynitride interface increases. Therefore, the location (and/or the density) of the defects responsible for the negative charge is changing during the deposition of the layer.

For the composition O/(O+N)=0.58 the charge distribution was studied after various anneal treatments. After an N_2 anneal (1000°C, 1hr) the amount of positive near-interface charge remains nearly the same but the localized negative charge is no longer present, indicating that the defect related to the negative charge is connected with the presence of hydrogen (because the N_2 anneal lowers the hydrogen content of the layers) or is removed by structural rearrangements. A subsequent H_2 anneal (1000°C, 1hr) introduces positive charge with a decreasing density from the outer surface towards the oxynitride bulk. This indicates a relation between positive charge and hydrogen. An NH_3 anneal (900°C, 1hr) has almost no effect on the spatially-localized negative charge while the density of positive charge near the Si/oxynitride interface is significantly lowered.

In the experiments on multi-step deposited layers no charge at the interface of the subsequently deposited layers could be evidenced. The positive charge located near the Si/oxynitride interface is, therefore, directly related to the presence of the Si-substrate and is not due to a non-equilibrium situation at the start of the deposition. This means that the Si-surface imposes a certain structure (and defect type) in the near-interface region of the oxynitride layer.

The Si-oxynitride interface trap density of the as-deposited layers as a function of the O/(O+N) ratio shows a narrow transition region at an O/(O+N) ratio of 0.35 where the interface trap density suddenly drops by almost two orders of magnitude. Below this O/(O+N) ratio a peak is found in the interface trap density typically located around 0.17 eV above the middle of the Si-bandgap. This O/(O+N) ratio of 0.35 is also a transition point for the annealing behaviour of the interface traps.

When a 30 nm oxide layer is present in-between the Si-substrate and the oxynitride layer, the interface trap density has a value typical for good thermal oxides. A 3 nm oxide barrier results in interface trap densities with values between these found for thermal oxides and for the oxynitride/Si-interface.

As the oxygen content of the layers increases, the conductivity and the charge trapping decrease. From Weinberg-measurements on transistor structures it was concluded that for O/(O+N) ratios up to 0.43 (larger O/(O+N) ratios were not measured) the conduction in the oxynitride layers is dominated by the hole current. The retention of the layers improves for O/(O+N) ratios larger than 0.35, demonstrating that it is closely linked to the interface trap density in MNOS capacitors.

Tests of the dielectric integrity of the layers revealed an increasing number of low-field breakdowns for increasing oxygen content of the oxynitride layers. The relatively low number of low-field breakdowns in the films with a low O/(O+N) ratio is related to the high trapping rate in these layers.

In chapter 5 the observations made on the electrical characteristics of the layers, reported in this chapter, will be correlated with the measurements on the physical properties, as reported in the other chapters. From this correlation a model will be proposed which relates the various properties to the structural details of the layers.

REFERENCES

1) H.E. Maes and J. Remmerie in "Symposium on Silicon Nitride Thin Insulating Films", (Electrochemical Society, Princeton, NJ), Eds. V.J. Kapoor and H.J. Stein, Vol. 83-3, p. 73 (1983)

2) E.H. Nicollian and J.R. Brews, "MOS (Metal Oxide Semiconductor) Physics and Technology", Wiley, New York (1982)

3) P. Balk in "Proceedings of the Electrochemical Society Fall Meeting", (Electrochemical Society, Buffalo, NY), p. 29 (1965)

4) S. Kar and W.E. Dahlke, Solid State Electron 15, 221 (1972)

5) G. Schols and H.E. Maes in "Symposium on Silicon Nitride Thin Insulating Films", (Electrochemical Society, Princeton, NJ), Eds. V.J. Kapoor and H.J. Stein, Vol. 83-3, p. 94 (1983)

6) E.H. Pointdexter, P.J. Caplan, B.E. Deal and R.R. Razouk, J. Appl. Phys. 52, 879 (1981)

7) P.M. Lenahan and P.V. Dressendorfer, J. Appl. Phys. 54, 1457 (1983)

8) P.M. Lenahan and P.V. Dressendorfer, J. Appl. Phys. 55, 3495 (1984)

9) E.H. Pointdexter, G.J. Gerardi, M.-E. Rueckel, P.J. Caplan, N.M. Johnson and D.K. Biegelsen, J. Appl. Phys. 56, 2844 (1984)

10) F.H.P.M. Habraken in "Insulating Films On Semiconductors", Eds. G. Declerck and R. De Keersmaecker, (North Holland, Amsterdam), p. 186 (1987)

11) J. Remmerie and H.E. Maes in "Insulating Films On Semiconductors", Eds. J.J. Simonne and J. Buxo (North-Holland, Amsterdam), p. 15 (1985)

12) Z. Weinberg, W. Johnson and M. Lampert, Appl. Phys. Lett. 25, 42 (1974)

13) J. Remmerie, H.E. Maes, J. Witters and W. Buellens in "Silicon Nitride and Silicon Dioxide Thin Insulating Films", Eds. V.J. Kapoor and K.T. Hankins (The Electrochemical Society, Pennington, NJ), p. 93 (1987)

14) E. Suzuki and Y. Hayashi, J. Appl. Phys. 53, 8880 (1982)

15) H.E. Maes and G. Heyns in "Insulating Films On Semiconductors", Eds. J. Verwey and D. Wolters (North-Holland, Amsterdam), p. 215 (1983)

16) F. Liou and S. Chen, IEEE Trans. Electron. Dev. ED-31, 1736 (1984)

17) D.K. Schroder and M.H. White, IEEE Trans. Electron. Dev. ED-29, 617 (1976)

Chapter 5

ON THE CORRELATION BETWEEN THE ELECTRICAL AND PHYSICO CHEMICAL PROPERTIES OF LPCVD SILICON OXYNITRIDE FILMS

F.H.P.M. Habraken[*], M. Heyns[x], H.E. Maes[x], R. de Keersmaecker[x], A.E.T.Kuiper[+] and W.F. van der Weg[*]

[*]Department of Atomic and Interface Physics, University of Utrecht, P.O. Box 80.000, 3508 TA Utrecht, The Netherlands.
[x] Interuniversity Microelectronics Center, Kapeldreef 75, 3030 Leuven, Belgium
[+] Philips Research Laboratories, 5600 JA Eindhoven, The Netherlands.

Abstract

In this chapter it is attempted to link the observations in the physico chemical part of the project with the results of the electrical measurements. Especially, the charge distribution in the as deposited and annealed LPCVD silicon oxynitride films will be discussed.

I. INTRODUCTION

In the preceding chapters results of a detailed experimental investigation of the physical and electrical properties of LPCVD silicon oxynitrides are presented and discussed. It is the aim of this chapter to link the results of the two types of analyses in order to enhance our understanding of the oxynitrides. This is not an easy and straightforward task: the electrical properties of insulators like silicon oxynitrides , are supposed to be determined for a large part by defects (minority sites) in the materials, whose concentration does probably not exceed a concentration of about 10^{20} cm^{-3}, whereas most mentioned physico-chemical characterization methods probe majority sites. This is the reason why in this study so much attention has been paid to the behaviour of hydrogen and chlorine in the oxynitrides. Hydrogen and chlorine are considered to represent or to cause deviations from the amorphous lattice structure and may therefore be responsible or indicative for electrically active defects. Their presence and behaviour during processing may give indications about local phenomena in the material, which are not easy accessible in another manner.

In the following section the results of the electrical and physico-chemical analyses are briefly summarized. We divide the properties in two categories. The first category envelops those properties which are expected to be dominated by local effects; in the other category are those properties which are merely an expression of the transition from silicon nitride to silicon dioxide upon increasing the O/O+N ratio and, therefore, are due to the average composition and structure of the materials. Afterwards a model will be presented in which some of the electrical properties and the behaviour of hydrogen can be discussed.

II SUMMARY OF EXPERIMENTAL RESULTS

In this section a summary is given of the results of the elctrical and physico-chemical characterization of LPCVD silicon oxynitrides. The results are divided in two categories A and B and are given in table 1 and table 2, respectively:

A) this category contains material properties which vary in a uni-directional way with varying O/O+N ratio from nitride to oxide (table 1)·

B) this category contains material properties which appear to vary in a more complicated way with the variation in oxygen concentration in the oxynitrides (table 2).

The data in the tables lend support for the view that the properties, ascribed to the average structure and composition (majority sites) vary uni-directionally from nitride to oxide.

TABLE I: CATEGORY A

Property	Behaviour with increasing R=O/(O+N)	Remarks
stress	tensile, decreases with increasing R	see ch. 1
refractive index	decreases with increasing R	see ch. 1
Cl concent.	increases with increasing R	see ch. 1
N-H IR peak position	increases with increasing R	see ch. 1
Auger param. Si(KLL)-Si(2s)	decreases with increasing R	see ch. 2
barrier height	increases with increasing R	see ch. 4
positive int. charge	constant with varying R	see ch. 4
permittivity	decreases with increasing R	see ch. 4

On the other hand, the electrical properties, ascribed to minority sites, show a peculiar behaviour at an O/O+N ratio (R) of approximately 0.35. Furthermore, two physico-chemical parameters show a peculiar behaviour at R≈0.35: the hydrogen concentration and the relative density. One working hypothesis might be that the presence and the behaviour of hydrogen, representing a minoriy site itself, determines the electrical properties of the material *or* that the hydrogen chemistry is determined by the same but unknown parameter in the material as the electrical properties are. This is further corroborated by the observation that the flatband band voltage increases as a result of N_2 annealing, which also results in a loss of hydrogen, but decreases again as a result of a subsequent H_2 anneal, which in its turn leads to a reintroduction of hydrogen. Probably also other parameters vary as a result of the heat treatment itself, but it is not clear why this variation is reversible with variation of the anneal ambient. For these reasons it is worthwhile to discuss the behaviour of hydrogen in more detail, because this may be the key to the understanding of the correlation between the electrical and physico-chemical properties.

Secondly, the density, relative to the crystalline constituents of the material, shows a maximum at R ≈ 0.35. This maximum persists after N_2

TABLE II: CATEGORY B

Property	Behaviour with increasing R=O/(O+N)	Remarks
H conc. as depo	constant for R<0.35 decreases for R>0.35	see ch. 1
H conc. annealed	maximum at R ≈ 0.35	see ch. 1
H uptake during H$_2$ anneal	minimal at R ≈ 0.35	see ch. 1
Relative density	maximum at R ≈ 0.35	see ch. 1
bulk charge as deposited	charge free for R<0.35 negative charge for R>0.35	see ch. 4
interface state density	decreases strongly at R≈0.35 increases upon annealing R<0.35 decreases upon annealing R>0.35	see ch. 4
conductivity	decreases strongly around R≈0.3	see ch. 4

annealing, although the hydrogen concentration shows a clear maximum at that composition after annealing and a slight maximum before the anneal. This is surprising because one might think that a maximum H concentration should result in a minimum in the relative density.

III DISCUSSION

According to the foregoing, the correlation between the electrical and the physico-chemical properties of silicon oxynitrides might be found in the interplay between the defect and hydrogen chemistry in the films in connection with their local structure. From the measurements of the flatband voltages in the as deposited and heat treated films we deduce that charge trapped in the oxynitride insulator can be positive and negative at the same O/O+N ratio.

In Si_3N_4 charge trapping effects have been investigated extensively. In accordance with the theoretical result of Robertson and Powell [1] that Si dangling bonds create states at midgap in silicon nitride, Fujita and Sasaki [2] observed a correlation between the concentration of memory traps in MNOS structures and paramagnetic centers, which they attributed to Si dangling bonds. Krick et al [3] demonstrate the existence of a strong correlation between changes in the density of paramagnetic silicon dangling bond centers and changes in the density of the space charge in amorphous silicon nitride films induced by alternate illumination and positive and negative charge injection. This latter study strongly indicates the amphoteric character of the Si dangling bond defect and its responsibility for the charge trapping behaviour of silicon nitride.

The trapped electron density as a result of internal photoemission in MNOS structures appeared to correlate with the amount of oxygen impurity in Si_3N_4 [4], with the amount of excess silicon [5] and with the Si-H bond concentration in the nitride [6]. From these observations we tend to conclude that the extent of charge trapping increases with increasing deviation from the Si_3N_4 stoichiometry. However, introduction of oxygen in the nitride to grow oxynitrides, as is the case in this work, does not result in an increase in trap density; on the contrary, there are strong indications that the trap density decreases with increasing oxygen concentration [7]. This indicates that the precise way of introduction of oxygen or its concentration determines whether it contributes to trap formation or not.

Notwithstanding the above mentioned indications that Si dangling bonds are responsible for the charge trapping behaviour in Si_3N_4 under charge injection and internal photoemission, we have to consider various kinds of defects in that material. Robertson and Powell [1] calculated neutral nitrogen dangling bonds states N^o to be within the valence band and the negatively charged N^- states to be positioned in the nitride gap slightly above the valence band edge. The neutral silicon dangling bond related center Si^o lies around midgap, having a positive electron correlation energy U. On the basis of their calculations the N^- states are responsible for the dominant hole conduction in silicon nitride, as is also observed in this study. Lucovsky and Lin [8] calculated the Si^o states to lie just below the conduction band edge of silicon nitride and oxide whereas the Si-Si native defect is around midgap. Insertion of a more electronegative element like oxygen in the nitride shifts the energy of the defects states a few tenths of an eV upwards whereas the opposite is the case for insertion of a less electronegative element like hydrogen [8]. Si-H defects have a level in the valence band [1] or a level 1 to 2 eV above the valence band edge [8]. Si-N-H defects do not produce states within the gap [8]. Finally, we must mention that recently Krick et al [9] argue that the correlation energy of the Si dangling bond related defect is negative. This has also been conjectured a few years ago by Ngai and Hsia [10].

IV THE MODEL

We start with the presupposition that in the insulator/semiconductor structure no net charge is present. In a situation, where an insulator/semiconductor structure is able to reach thermal equilibrium with respect to its charge distribution, positive charge is expected in the insulator, if defect levels in the insulator are positioned above the semiconductor Fermi level, which are neutral when filled with electrons. In this situation these defects will be emptied and electrons will transfer from the insulator to the semiconductor. This will occur at an appreciable rate at high temperatures, for instance during deposition of the insulator or annealing of the layer structure. During cool down of the sample the high temperature charge distribution will be frozen in. In contrast, during the same treatments, electrons will transfer from the semiconductor to the insulator if insulator empty states are located below the semiconductor Fermi level. This results in the build up of negative charge in the insulator. To what extent the charge transfer will occur depends on the concentration of the defects and also on the relative trapping and diffusion rates of free holes and electrons in the interface region [11], which on its turn influences the electric field distribution in the system.

Since in the Si/oxynitride system positive or negative charge has been found to be present, without non-equilibrium injection into the layer, it is concluded that both low lying (Lo states) and high lying defect levels (Hi states) can be present. Then the amount and sign of the stored charge depends on the relative concentration of the considered two type of defect states. Under thermal equilibrium no charge will be stored in the insulator if the concentration of these low lying states and the concentration of the high lying states are equal. This apparently is the case in the as deposited LPCVD silicon (oxy)nitrides having $O/(O+N)<0.35$. In these materials compensation of positive and negative charge exists, which we believe not to be fortuitous and which effect will be discussed later. Negative charge is induced in (oxy)nitrides with $O/(O+N)<0.35$ during $1000°C$ annealing, which indicates the existence of an excess of low lying states in these materials. A subsequent H_2 anneal results in a decrease in the amount of negative charge stored. Therefore the concentration of low lying empty states has decreased or the concentration of high lying states has increased. During N_2 annealing the largest physico-chemical effect is the decrease in the N-H bond concentration;

during subsequent H$_2$ annealing the N-H bond concentration increases again. In view of these observations we interpret a part of the low lying states to be associated with N dangling bonds, which are therefore negatively charged. This assignment is in agreement with the calculations of Robertson and Powell [1]. The high lying states are supposed to be associated with Si dangling bonds with their amphoteric character.

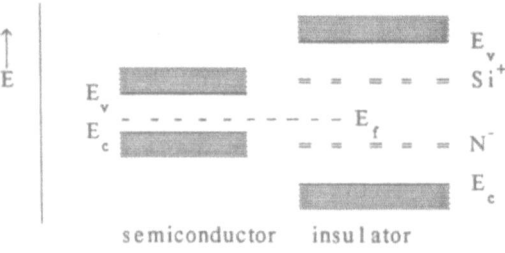

Fig. 1. Schematic band diagram of a semiconductor/insulator structure.

For O/(O+N)>0.35 negative charge is already present in the as deposited silicon oxynitrides. Within the above given picture this indicates that the concentration of low lying states is larger than the number of high lying states. Furthermore, for O/(O+N) > 0.35 the amount and sign of stored charge does not parallel the loss and reintroduction of hydrogen. In contrast, during annealing an increasingly larger amount of positive charge is stored in the oxynitrides, irrespective of the anneal ambient. Furthermore, the charge is not distributed homogeneously over the bulk of the oxynitride films. The centroid of the negative charge in the as deposited material moves away from the interface at larger oxynitride thicknesses whereas the positive charge in the first hour of (N$_2$) annealing at 1000°C appears in the interface region and in the second hour of 1000°C (H$_2$) annealing it distributes over the bulk of the film. This indicates that the oxynitride film is not in equilibrium. This may have two causes: the concentration and the nature of the defects vary during deposition (at 800°C) and annealing (at 1000°C) or the occupation of the levels varies during these high temperature treatments. Arguments in favor of the latter explanation might be the opening of the gap for O/(O+N)>0.35 as found for plasma oxynitride films by Michailos and Cros [12] and/or the strong decrease of the interface state density at this composition (see chapter IV). Both effects may make charge transfer from the substrate to the insulator more difficult for larger oxygen contents in the oxynitrides. This is also illustrated by the transition of the Poole-Frenkel type of conduction to a Fowler-Nordheim type of conduction (ch. IV). This effect may be further reinforced by the observation that the deep defect density in the insulator decreases with increasing oxygen concentration.

At the interface positive charge is present already in the as deposited samples. This amount of interfacial charge appears to be constant after the anneal treatments. This certainly finds its cause in the presence of the interface and probably, more specifically, in the presence of the native oxide on the c-Si substrate. In chapter I it is argued that this native oxide is converted into an oxynitride through the reaction of the oxide with NH_3. It is known that thermal nitridation in SiO_2 using NH_3 leads to the incorporation of positive charge in the insulator, also when the nitridation is carried out at low ammonia pressures [13]. The positive charge build-up in the annealed LPCVD silicon oxynitrides having the larger O/(O+N) ratios, may have the same origin as the positive charge generated during thermal nitridation of silicon dioxide films.

A detailed understanding of the electrical behaviour of the oxynitride films involves probably an understanding of the role of the incorporated Cl, which is presumed to be negatively charged, and the O/(O+N) dependence of the position of the Si, N and O dangling bond associated levels relative to the semiconductor Fermi level. Furthermore, probably an interrelation between the existence of the different types of defects exists, including the formation of of Si dangling bonds originating from Si-Si native defects. This interrelation is probably the origin for the charge compensation at low O/(O+N) ratios, which compensation apparently breaks down at larger O/(O+N) ratios. The interrelation between the presence of Cl on the one hand and the concentration of hydrogen and therefore probably also the concentration of N and Si dangling bonds on the other hand has been indicated by the observation of the accelerated removal of H from oxynitrides during heat treatment in a Cl containing ambient.

REFERENCES

[1].J. Robertson and M.J. Powell, Appl. Phys. Lett. 44, 415 (1984).

[2]. S. Fujita and A. Sasaki, J. Electrochem. Soc. 132, 398 (1985).

[3]. D.T. Krick, P.M. Lenahan and J. Kanicki, J. Appl. Phys. 64, 3558 (1988).

[4]. V.J. Kapoor, R.S. Bailey and S.R. Smith, J. Vac. Sci. & Technol. 18, 305 (1981).

[5]. R.S. Bailey and V.J. Kapoor, J. Vac. Sci. & Technol. 20, 484 (1982)

[6]. V.J. Kapoor, R.S. Bailey and H.J. Stein, J. Vac. Sci. & Technol. A1, 600 (1983).

[7]. Ch. IV. See also: J. Remmerie, Thesis University of Leuven (1987).

[8]. G. Lucovsky and S.Y. Lin, J. Vac. Sci. & Technol. B3, 1122 (1985).

[9]. P.M. Lenahan, D.T. Krick and J. Kanicki, Appl. Surface Sci. 39, 392 (1989).

[10]. K.L. Ngai and Y. Hsia, Appl. Phys. Lett. 41, 159 (1982).

[11]. R.C. Hughes, J. Appl. Phys. 56, 1044 (1984).

[12]. J. Michailos, thesis Université Joseph Fourier, Grenoble, 1989.

[13]. W. Yang, R. Jayaraman and C.G. Sodini, IEEE Trans. Electron Devices 35, 935 (1988).

Chapter 6

The use of oxynitride layers in non-volatile S-OxN-OS (silicon-oxynitride-oxide-silicon) memory devices

Herman E. Maes

Interuniversity Microelectronics Center, Kapeldreef 75,
B-3030 Leuven, Belgium

ABSTRACT

The memory characteristics of non-volatile S-OxN-OS (silicon-oxynitride-oxide-silicon) memory transistors have been studied for oxynitride compositions in the range of $[O]/[N] = 0$ (pure nitride) to 0.75.

The main observations from this study are : the decreasing trap density in oxynitride layers with increasing oxygen content, the decreasing leakage of stored charge out of the layer for equal amounts of stored charge resulting in a constant threshold voltage decay rate for equal amounts of stored charge or a decreasing decay rate for an equal memory window and the enhanced endurance for oxynitride structures with low oxygen content.

From this study it is found that the optimum composition for the non-volatile memory function is obtained for a value of $[O]/([O]+[N])$ between 0.1 and 0.2 in view of the better retention and endurance characteristics.

1. Introduction

MNOS (metal-nitride-oxide-silicon) non-volatile memory devices have been studied extensively in the past but these studies were almost merely carried out on devices with silicon nitride as the charge storage medium. These nitride layers were obtained by deposition either at atmospheric pressure (APCVD) or at low pressure (LPCVD). Only recently studies have been devoted to the use of LPCVD oxynitride layers in simple capacitor or transistor structures [1,2]. The use of LPCVD oxynitrides in MNOS circuits however has not yet been reported until now. Combinations of a nitride layer and an oxide layer in the so-called SONOS structures on the other hand have received considerably more attention [3-8]. Oxynitrides nevertheless show properties that make them a viable and often preferable alternative for the pure nitride film in memory device applications. An extensive study was therefore conducted by us on the use of oxynitride layers (referred to as OxN) with various compositions ranging from the pure silicon nitride to layers with a [O]/[N] ratio of 0.75. The purpose of this paper is to report on the memory characteristics of these transistors, i.e. on their programming, retention and endurance behaviour. Although a detailed experimental studied was also carried out previously on the transport properties of the same oxynitride layers as compared to those of pure nitride layers [9] and on the properties of triple layer devices (O-OxN-O) [2], this report will only deal with the mentioned memory properties of S-OxN-O-S transistors.

2. Device fabrication

The devices were fabricated in a 5 μm nMOS polysilicon-nitride-oxide-silicon process. The memory devices consist of a 2 nm thick oxide obtained by thermal growth at 650°C during 15 min in dry oxygen. The oxynitrides were deposited at 820°C in an LPCVD reactor using a 3:1 mixture of NH_3 + N_2O:SiH_2Cl_2. The composition of the films was varied between pure Si_3N_4 and a composition yielding a [O]/[N] value of 0.75 by controlling the NH_3/N_2O ratio. The total (NH_3 + N_2O) flow was kept constant at 75 sccm. In total 5 different compositions were obtained. The oxynitride layer thickness varied between 34-35 nm. All information on the different gate insulator layers such as nature of the layer, structure, used gas flows, value of x= [O]/([O]+[N]), layer thickness, dielectric constant and refractive index are given in Table 1. The dielectric constants were obtained from capacitor-voltage measurements and the x-values were obtained from RBS and IR absorption measurements [10-11]. During the reflow of the contact hole oxide layer, a hydrogen anneal was done at 800°C for 15 min in order to reduce the interface trap densities in the transistors. Complete physical and electrical characterization of these oxynitride layers has been performed by all partners in the ESPRIT 369 project and is reported in ref [10].

structure	d_{on} (nm)	NH_3/N_2O	ε_n	x = [O]/([O]+[N])	n
nitride	34	75/0	7.00	0.00	2.00
oxynitride	34	48/27	6.55	0.14	1.88
oxynitride	35	40/35	6.25	0.20	1.83
oxynitride	35	30/45	5.60	0.35	1.76
oxynitride	35	20/55	5.25	0.43	1.66

Table 1   Information on the different devices used in this study

3. Experimental results and discussion

The extensive set of experiments that have been carried out on these memory transistors include evaluation of the interface trap densities using the charge pumping technique [12-13], determination of the programming characteristics, retention measurements and endurance behaviour. Before describing these results in more detail, first some characteristics of these oxynitride layers that will be needed in the further discussions are shown in the first 5 figures. These characteristics are all shown versus the x = [O]/([O]+[N]) ratio. Figure 1 shows the reduction of the refractive index and the dielectric constant of these oxynitrides with increasing oxygen content (also listed

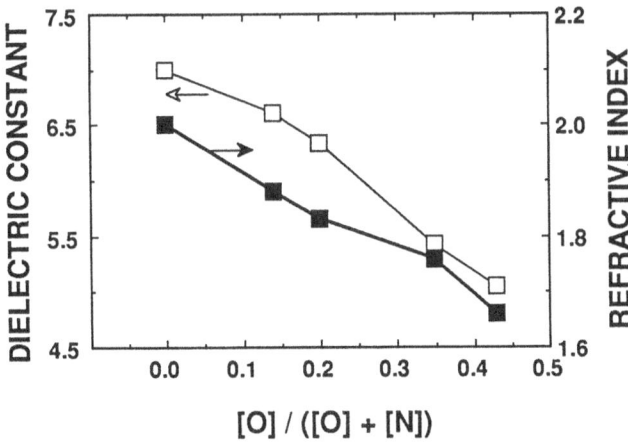

Figure 1. Dielectric constant and refractive index of oxynitride layers as a function of [0]/([0]+[N])

in Table 1). The relative dielectric constant drops from a value of 7 for the pure nitride to 5.25 for the layer with x=0.43. In a previous study on oxynitride layers [14] a thorough evaluation was made of the conduction and trapping in these layers. The most important parameter from that study for the purpose of this report is the threshold injection field. This threshold field is defined as the electric field at the injection site for which significant trapping starts to occur in the oxynitride layer when slowly ramping the voltage over the oxynitride layer. When monitoring the flatband voltage shifts in these layers during positive gate voltage ramping, an abrupt increase of the V_{FB} shifts is observed [14]. Figure 2 shows the injection field for injection from the silicon substrate. As can be seen this field increases significantly with increasing oxygen content in the layer. This

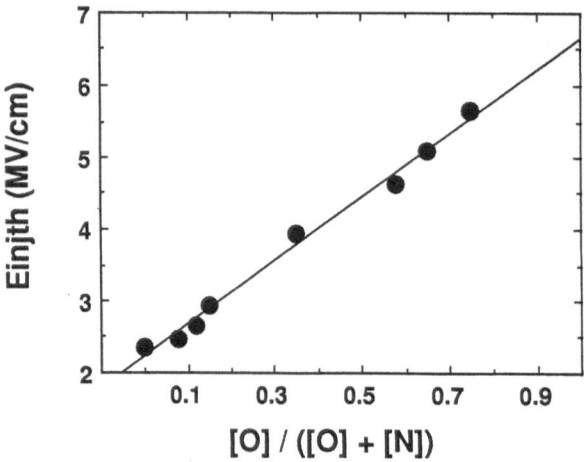

Figure 2. Injection field at the onset of charge trapping in oxynitride layers as a function of [O]/([O]+[N])

can be understood in terms of the expected increasing energy barrier at the silicon-oxynitride interface for layers with increasing oxygen content. In order to get an idea of this energy barrier we started from the expression for the Fowler-Nordheim tunneling current [15] and assumed a same injection current level for all layers at the abrupt onset of trapping. This means that in a first approximation

$$\Phi_B^{3/2} / E_{inj} = \text{constant} \tag{1}$$

for all layers, with Φ_B the energy barrier and E_{inj} the injection field shown in figure 2. Using eq (1) and assuming that for the pure nitride (x=0) Φ_B=2 eV, the values for the barrier height of all oxynitride layers can be obtained. This was done for the data of figure 2. The results are shown on figure 3. It is clear from this figure that the barrier height increases with oxygen content as expected. The results obtained through eq (1) are however only a rough approximation. For a pure oxide the barrier height is expected to be around 3.2 eV. From this first approximation however an extrapolated value of about 3.9 eV is obtained.

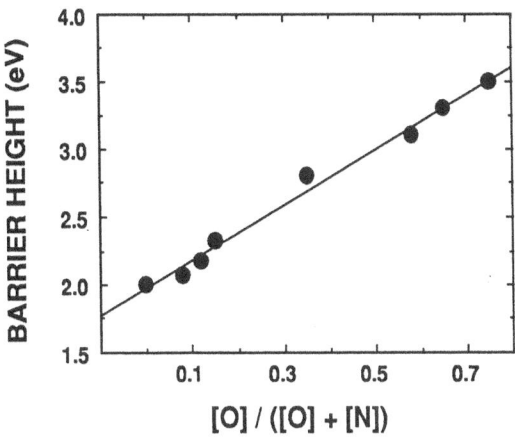

Figure 3. Barrier height at the Si-OxN interface for oxynitride layers as a function of [O]/([O]+[N]), computed from the data of figure 2 and eq (1)

From the study of the charge content and distribution of as-deposited oxynitride layers in large capacitors [10] the presence of positive and negative charge was evidenced as is shown on figure 4. In this figure the flatband voltage is shown as a function of the oxynitride layer thickness. This

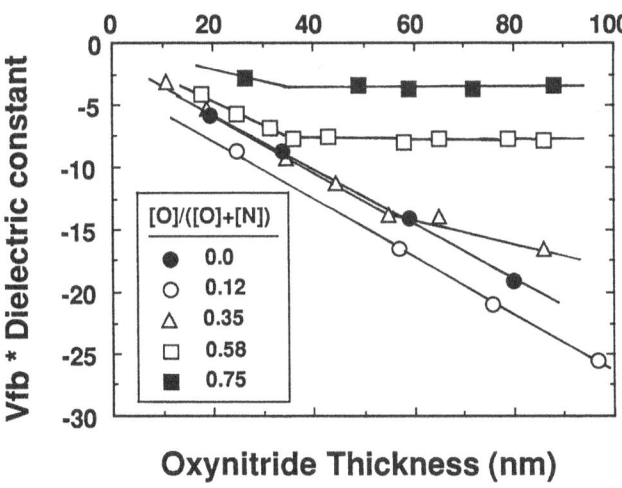

Figure 4. Variation of the Flatband voltage with thickness of oxynitride layers with different values of [O]/([O]+[N])

was obtained by backetching of the oxynitride and by measuring the corresponding flatband voltage. From this figure it was concluded first of all that a large positive charge is present at the Si-OxN interface which however decreases for x ≥ 0.4 and secondly that a negative charge builds up for increasing [O] content with a centroid located at some distance from the interface [10]. This is clearly seen on figure 4. In reference [10] it was also shown that the interface trap density measured on these capacitors using the CV-technique is high for layers with low [O] content but drops for x > 0.5. This is consistent with the observation that the positive charge is reduced for higher oxygen content. Therefore it is believed that the positive charge located at the interface is identical to or closely related to the interface trap charge. In figure 5 the threshold voltage (V_{th}) of the transistors recorded immediately after fabrication and prior to any other characterization is shown. It is seen that this initial V_{th} is increasing with increasing oxygen content. For moderate x-values the increase is linear but for larger values (x>0.35) a sudden increase is noticed.

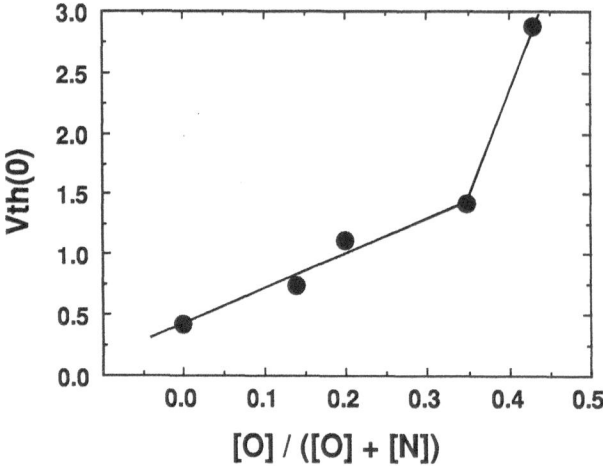

Figure 5. Threshold voltage of the virgin S-OxN-OS transistors as a function of [O]/([O]+[N])

3.a The interface trap density

The interface trap density in the S-OxN-OS and MOS transistors on the chip are shown in figure 6 as a function of layer composition. Whereas for MOS transistors a low interface trap density ($< 2x10^{10}$ cm^{-2}eV^{-1}) is to be expected, these low densities on SNOS type of devices can only be obtained by an appropriate hydrogen anneal [16]. The effectiveness of this step is clear from figure 6. First of all the measured values are of the order of $2x10^{10}$ cm^{-2}eV^{-1} and secondly these values are independent of composition which is expected for the used

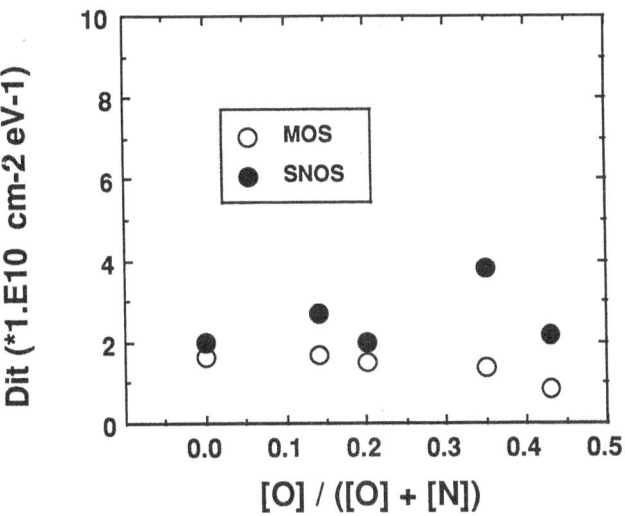

Figure 6. Interface trap density in MOS and SNOS devices as a function of the [O]/([O]+[N]) value of the oxynitride layer on that wafer.

hydrogen anneal. This is in contrast with the results from CV-measurements on the S-OxN-OS capacitors mentioned above [10]. The values of the latter are 1 to 2 orders of magnitude higher and strongly composition dependent. On these capacitors no hydrogen anneal was carried out and such an anneal would anyhow not have been very effective for the used temperature (800 °C) and time (15 min) [10,17]. In view of the low interface trap densities measured on the S-OxN-OS transistors, the result and behaviour of figure 5 can be understood. The negative charge that builds up for increasing oxygen content in the layers as evidenced from figure 5 is also expected to be present in the transistors. The large positive charge that was present in the capacitor structures is however not present in these devices. Therefore the increase of the threshold voltage with increasing oxygen content can be directly related to the larger negative charge. As was also evident from figure 5, this charge increases strongly for x > 0.35.

3.b Transient behaviour

The programming characteristics of the memory transistors were obtained by applying a positive voltage (write operation) or a negative voltage (erase operation) of *equal amplitude* at the gate of the transistors while grounding source, drain and substrate. A typical example is shown in figure 7 for the case of the pure nitride layer. The initial threshold voltage of the device was that corresponding to the center of window (see further), in this case equal to -0.2 V. As is clear from this figure and as was found for all layer compositions, the erasing process (hole injection) is a much slower process than the writing process (electron injection). This is in agreement with all

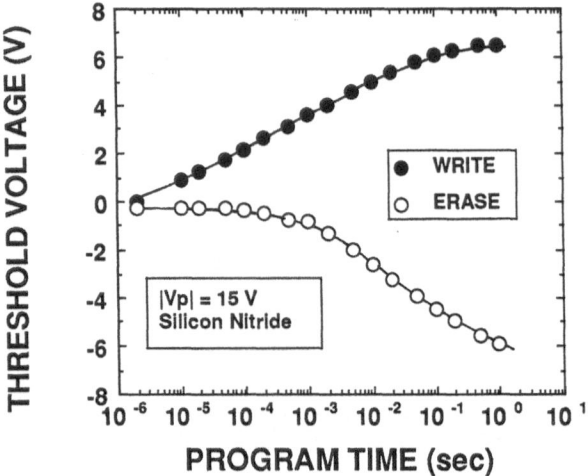

Figure 7. Transient characteristics of the transistor with a pure silicon nitride as the storage insulator. Program voltage amplitude is 15 Volt.

previous findings on MNOS devices. Typically, for the same voltage amplitudes, the erase action shows a delay of slightly more than 1 decade in time as compared to the write operation. Therefore in this comparative study it was decided to use programming conditions with equal amplitude for the erase and write voltages but with write pulses of 10 msec and erase pulses of 100 msec width. The purpose was of course to obtain symmetrical threshold voltage shifts for written and erased states. For the endurance evaluations however shorter pulses were used (but still preserving a ratio of 10 for erase over write time). When recording the transient characteristics up to saturation in the two states (in practice up to 1 sec (write) - 10 sec (erase), see figure 7) it was found that the center of the thus obtained saturation memory window is strongly dependent on the insulator composition. This center of the saturation window is shown on figure 8 as a function of x. Again, this value is indicative of fixed charges present in the insulator. This can also be concluded from a comparison with the initial threshold voltage values shown on figure 5. Whereas the center of window is slightly negative for the pure nitride layer, it becomes strongly positive for a high oxygen content in the layers. When using the write and erase pulse widths described above and equal amplitudes for both programming operations, the center of the thus obtained memory window will almost coincide with the center of saturated window shown in figure 8. So, for a value of x of about 0.14 the center of window will be zero, which means that in this case and for the chosen pulse width, the absolute values of the threshold voltage in the written and erased states are equal.

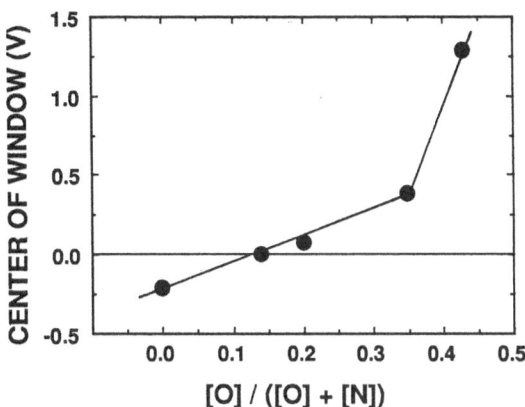

Figure 8. Center of window (i.e. the difference between the threshold voltages in the saturated written and erased states) as a function of the [O]/([O]+[N]) value of the oxynitride layer.

3c. Programming voltages

In a first experiment the voltages were determined which are required to obtain a 4V shift of the threshold voltage for the selected pulse widths starting from a virgin device (initial threshold voltage shown in figure 5). In figure 9 the measured values recorded for 10 msec write pulses are

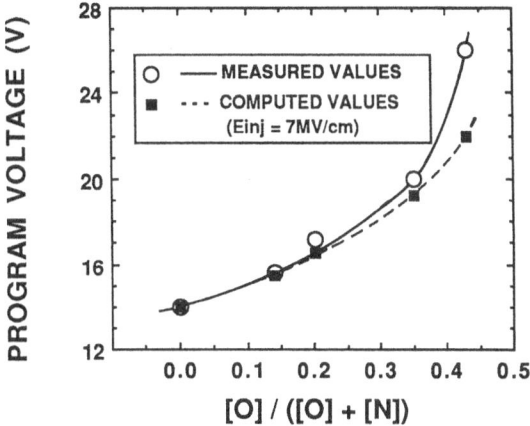

Figure 9. Program voltage (-O-; write operation) required in order to induce a 4V shift of the threshold voltage with a 10 msec pulse starting from the virgin state of the device. The computed values (-■-) are the gate voltages according to eq(2) required to obtain an initial injection field in the tunneling oxide of 7MV/cm for the different oxynitride layer compositions.

shown as a function of x. As can be seen the required program voltage increases strongly with increasing oxygen content. In order to understand the origin of this strong increase, we can first of all consider the relationship between the applied voltage (V_G) over the structure and the field in the tunnel oxide (E_{ox}), which in a first approximation is given by [15] :

$$V_G - V_{FB} = E_{ox} \cdot [d_{ox} + d_{on} \cdot \varepsilon_{ox}/\varepsilon_{on}] \qquad (2)$$

where V_{FB} is the flatband voltage, d_{ox} and d_{on} are the thickness of the tunnel oxide and the oxynitride layer respectively and ε_{ox} and ε_{on} are the dielectric constants in both layers. The flatband voltage at the start of the write operation can be obtained from the threshold voltage value $V_{th}(0)$ of the virgin devices shown in figure 5. For the used technology conditions (ϕ_F, depletion charge) we verified that :

$$V_{th}(0) \cong V_{FB}(0) + 1 \qquad (3)$$

From eq (2) it can be expected that the required program voltage V_G will indeed increase with increasing oxygen content of the layer, because of the decrease of the dielectric constant ε_{on} (see figure 1) and the increase of V_{FB} with increasing oxygen content according to eq (3) and figure 5. If we assume in a first approximation that an equal oxide field at the start of the write operation will result in an equal injection current and also similar transient behaviour for all compositions (this already implies that the trap density in oxynitride layers decreases with increasing oxygen content, see further), we can compute from eqs (2) and (3) and from figure 5, the required voltage for each composition, taking the device with the pure nitride as a reference. For this structure and from the measured 14V program voltage required to obtain a 4V threshold voltage shift for a 10 msec write pulse, we compute an initial oxide field of 7 MV/cm. Using this value as the initial oxide field for all other structures we obtain the values shown on figure 9. It is clear that although the increase of the required voltage can be rather well explained by this simple model for low oxygen contents, the deviation becomes significant for higher x values. The deviation could be partly explained by the fact that an equal oxide field does not correspond to equal injection conditions. This is shown in figure 10, based on the results for the barrier heights from figure 3, for the cases of a pure nitride ($\Phi = 1.1 eV$) and an oxynitride with x=0.43 ($\Phi = 0.5$ eV). In figure 10 the band diagrams are shown for the Si-SiO$_2$-SiOxN system for the injection conditions considered above (E_{ox} = 7 MV/cm, positive gate voltage). Image force lowering has been neglected. It is clear that in spite of the equal tunnel oxide field, the effective "injection barrier" for the modified Fowler-Nordheim current [15] is larger in the case of the oxynitride layer.

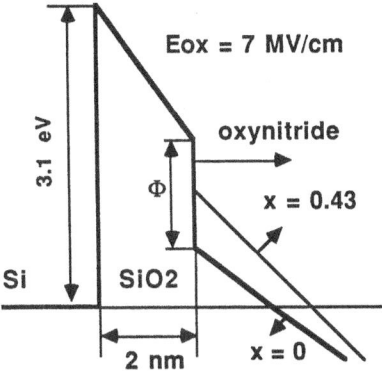

Figure 10. Band diagram of the Si-SiO$_2$-OxN system with a 7 MV/cm field in the tunneling oxide for two cases: pure nitride (x=0, ϕ=1.1 eV) and oxynitride (x=0.43, ϕ=0.5 eV)

The effect on the threshold voltage (or flatband voltage) of trapping of charges after injection is given by [15] :

$$\Delta V_{th} = - [\Delta Q_{on}/ \varepsilon_{on}].(d_{on} - x_c) \qquad (4)$$

in which ΔQ_{on} represents the trapped charge and x_c the centroid of this trapped charge distribution as measured from the SiO$_2$-SiOxN interface. As expected from eq (4) the decrease of the dielectric constant with increasing oxygen content should result in larger shifts of the threshold voltage for equal amounts of trapped charges in different oxynitride layers. An experiment was carried out in which devices with different oxynitride layer compositions were written into saturation with a voltage corresponding to an initial oxide field of 7.8 MV/cm for all structures. It was found that the saturation threshold voltages for the oxynitride layers with x up to 0.35 were all almost equal but were significantly lower (45%) for the layer with x=0.43. As was already discussed above the injection conditions for the latter layer are strongly different from those with lower x value. If we assume in a first approximation that all charge is stored close to the SiO$_2$/SiOxN interface ($x_c \sim 0$), the trapped charge in saturation can be obtained from the measured V$_{th}$ shifts using eq (4). This is shown on figure 11. The decrease of the trapped charge in the oxynitride layer with increasing x-value indicates that the trap density decreases with increasing oxygen content. In a first approximation and for moderate x-values this reduction in the trap density follows the ε_{on} reduction. The full line in figure 11 corresponds to a decrease proportional to ε_{on}. Anyhow, according to eq (4) a shift of 4V of the threshold voltage in the write experiment described above will correspond to lower trapped charge densities in the layers with larger oxygen content. Whereas in figure 9 the programming was performed starting from a virgin device, in a second

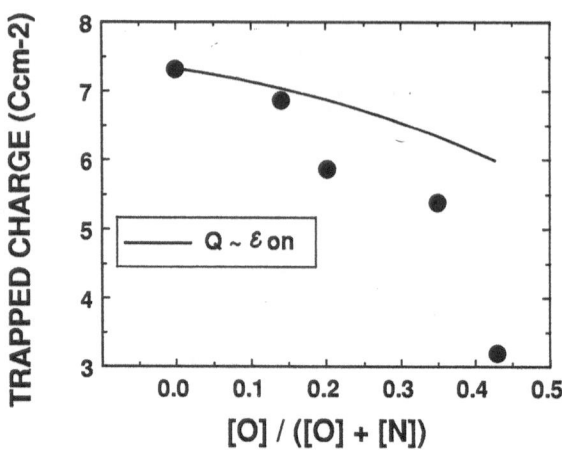

Figure 11. Trapped charge in saturation in oxynitride layers as a function of [0]/([0]+[N]) for equal initial injection conditions (E_{ox} = 7.8 MV/cm)

set of experiments the program voltage amplitude was determined for which, with the previously selected write and erase pulse widths, an effective window of 4V could be obtained. This is schematically shown in figure 12. These program voltage amplitudes are shown on figure 13 as

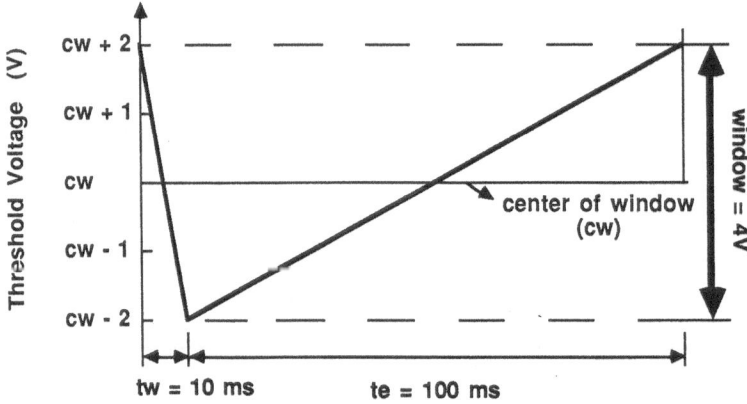

Figure 12. Programming sequence: 10 msec write pulses followed by 100 msec erase pulses. The threshold voltage window is equal to 4V. Definition of the center of window.

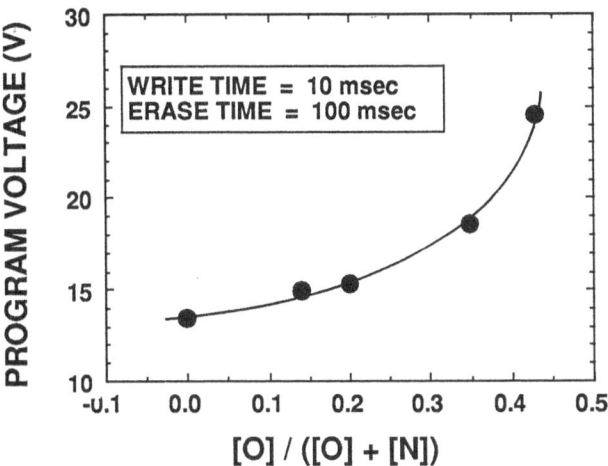

Figure 13. Program pulse amplitude according to the programming sequence of figure 12 as a function of the [0]/([0]+[N]) value of the oxynitride layer.

a function of x. It is again clear that these values increase strongly with oxygen content, essentially for the same reasons as discussed before for figure 9. The measured values are however slightly smaller (although in both cases the effective voltage shifts are 4V). This is however expected since the effect of the charge that was built up in the previous program operation will be to enhance the initial oxide field when changing the voltage polarity in the subsequent program operation, according to eq (2).

So from the experiments described in this section (figures 9 and 13) it can be concluded that the program voltages increase with increasing oxygen content but that this increase is moderate for low oxygen contents. In addition it was shown that the trap density in oxynitride layers decreases with increasing oxygen content.

3d. Retention behaviour

One of the most important characteristics of a non-volatile memory device is its information retention. It is known that MNOS devices do suffer a loss of their information starting almost immediately after the program operation when the gate is returned to zero potential. This information loss in both states is due either to backtunneling of trapped charges from the traps in the nitride layer towards the silicon [15,18] or to tunneling of carriers of the opposite sign from the silicon into the nitride layer and compensation of trapped charges [19] or a combination of both. In any case the reduction of the threshold voltage follows a logarithmic time dependence. The slope of this logarithmic dependence is known as the *decay rate* and is expressed in Volts/

(decade of time). Since tunneling is strongly field dependent, it can be expected that larger amounts of trapped charge will result in larger decay rates. Since in a first approximation the decay rates scale with the initial amount of trapped charge present [15,18] and consequently also approximately with the corresponding initial threshold voltage swing, it is common practice to define a relative decay rate which would then be the above defined decay rate divided by the initial threshold voltage, i.e. :

$$\text{Relative decay rate} = \Delta V_{th}/[(\text{decade of time})*V_{thin}] \qquad (5)$$

If one starts from an initially "charge free" insulator, the relative decay rate (for one state in eq (5)) is sufficient to predict the retention of the devices for all amounts of trapped charge in that state. These relative decay rates were determined for all S-OxN-OS devices programmed as described in the previous section, i.e. with a final memory window of 4V. Figure 14 shows the relative decay rates as a function of x. It can be seen that for the erased state these relative decay rates are almost composition independent but that they decrease significantly with oxygen content

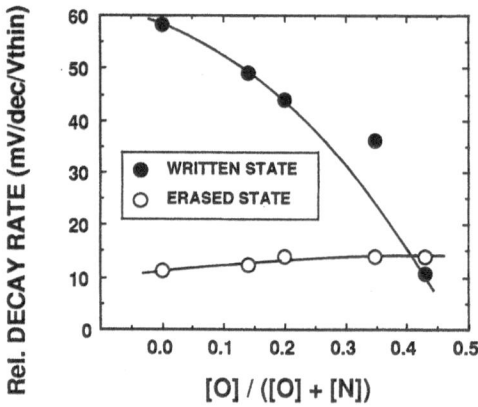

Figure 14. Reliability decay rate of the unstressed devices in written and erased state as a function of the [0]/([0]+[N]) value of the oxynitride layer.

for the written state. This latter effect is however partly due to the fact that for increasing oxygen content the initial threshold voltage in the written state is strongly increasing. This can indeed be understood with reference to the discussion in section 3b and the increase of the center of window as shown in figure 8. For example, for the pure nitride V_{thin} in the written state will be 1.75V whereas for x=0.35 this would already be 2.4V. As mentioned before, a shift of 2V in V_{th} will correspond to a lower amount of trapped charges for the layers with a larger oxygen content. This explains why, even after accounting for the larger initial threshold voltages for larger oxygen content, the decay rates are decreasing with x for the 4V memory windows used here. For the erased state the absolute value of the initial threshold voltage is decreasing with increasing x-value. This indicates that the absolute decay rate itself will therefore be decreasing with increasing oxygen content. In order to check how the decay rates vary in oxynitride layers if a same amount of charge is stored in the different devices, the following experiment was set up. The devices were written and erased using such programming conditions that :

$$\Delta Q = \Delta V_{th} \cdot \varepsilon_{on}/d_{on} \qquad (6)$$

was constant for all devices. The charge in the nitride layer was 3.6×10^{-7} Ccm^{-2} for a 2V shift in the written and in the erased state. In order to store the same amount of charge in the different oxynitride devices, considerably larger ΔV_{th}'s (according to eq (6)) were necessary and consequently also higher programming voltages than those predicted from figure 13. The decay rates measured in these devices with the same amount of stored chrage are shown in figure 15. As can

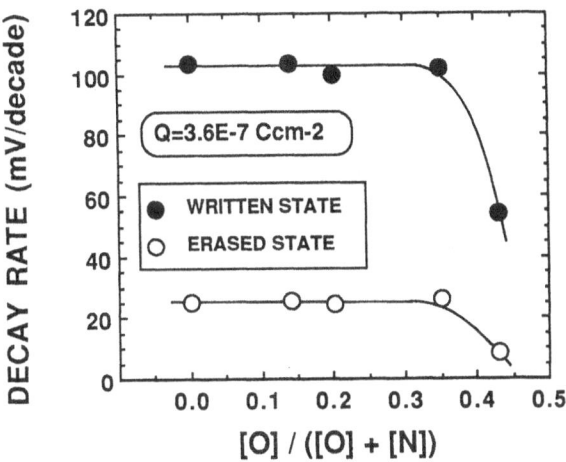

Figure 15. Decay rate of the unstressed devices containing an equal amount of stored charge, as a function of the [0]/([0]+[N]) value of the oxynitride layer.

be seen, up to x=0.35 the decay rates are almost constant but for larger x-values the decay rates are strongly reduced. The fact that the threshold voltage decay rate is constant means in fact that the decay of the stored charge is decreasing with increasing oxygen content, again almost in proportion to the decrease of the dielectric constant (see eq (6)). This is related to two facts, first of all the trap density is decreasing with increasing oxygen content : although equal amounts of charge have been considered, this charge is distributed deeper into the layer for larger oxygen content. Secondly, the energy depth of the traps is increasing with oxygen content. The drastic decrease in decay rates for the larger x-values can also be explained in these terms and in view of the strong decrease in trap density as evidenced from figure 11.

From this section we can conclude that the lower decay rates measured in devices with increasing oxygen content in the insulator for equal threshold voltage shifts are due to the lower amount of charge that needs to be stored to realize these shifts and to the reduced backtunneling of charges for equal amounts of stored charges in all these layers. Up to x=0.35 the decay rates for equal amounts of stored charge are almost the same in all devices. This is the case for trapped electrons as well as for trapped holes. Only for larger x-values these decay rates strongly decrease.

3e. Endurance behaviour

The degradation of the transistors is evaluated by using higher programming voltages and smaller pulse widths (100 μsec for write and 1 msec for erase). The voltages were however increased according to the values of figure 13 and the resulting memory window was still 4V. Figure 16 shows the increase of the interface trap density in the different devices as a function of the number of accumulated programming cycles. It is seen from this figure that the increase is faster for

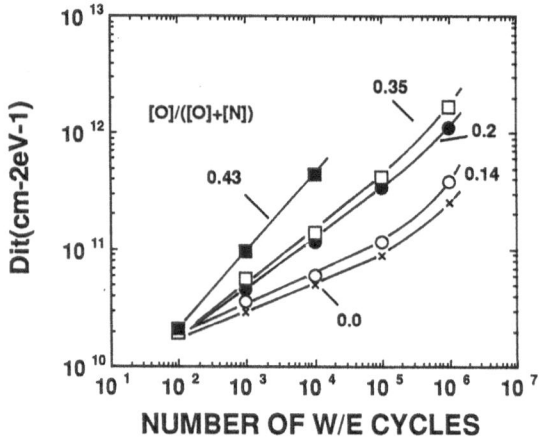

Figure 16. Increase of the interface trap density as a function of the accumulated number of Write/Erase cycles for transistors with different oxynitride compositions.

increasing x-values. Especially for values of x>0.2 this increase is significant. It is believed that this is partly due to the higher oxide fields at the start of erase of the devices with layers with larger oxygen content. As is known, hole transport through the thin oxide is the major cause for interface trap generation in MNOS devices [20]. For example for the nitride layer the field at the start of the erase operation (for the higher pulse amplitudes used here and the 4V window) is 9 MV/cm whereas for the device with x=0.35 this field is 10.3 MV/cm.

Figure 17 shows the shift of the center of window with cycling. For these higher stressing

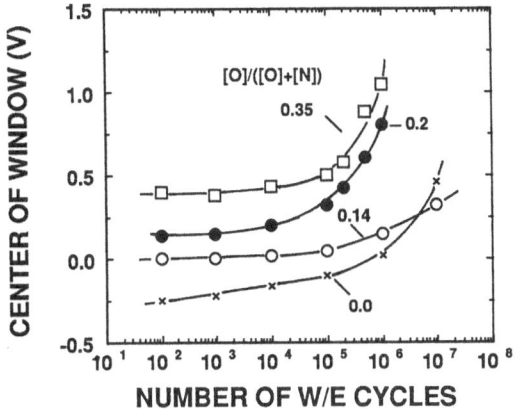

Figure 17. Shift of the center of window with cycling for transistors with different oxynitride compositions.

conditions the center value shifts only slightly for x=0.14, being less than 0.2V after 10^6 cycles. However for larger x-values the center of window increases strongly after 10^5 cycles showing a positive shift which is indicative for the build-up of negative fixed charge. Figure 18 again shows the relative decay rates as defined above as a function of x after accumulation of 10^6 cycles. The

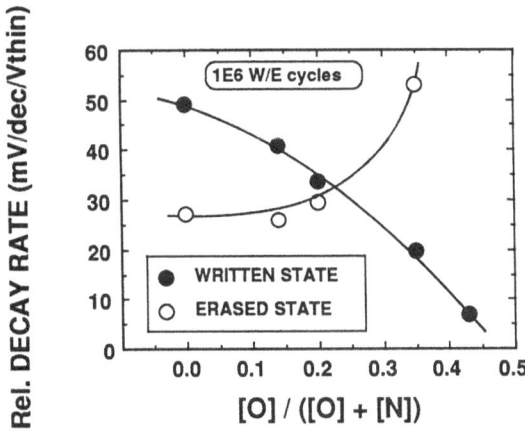

Figure 18. Relative decay rate of the stressed devices (10^6 cycles) in written and erased state as a function of the [O]/([O]+[N]) value of the oxynitride layer.

lower values on this figure for the written state but the higher values for the erased state can be understood in view of the positive shift of the center of window (larger positive V_{th} in the written state, lower absolute value of V_{th} in the erased state). In order to allow a fair comparison, figure 19 again shows the V_{th} decay rates for an equal amount of stored charge (same value as on figure 15) after 10^6 accumulated cycles. When comparing these results with those for the unstressed

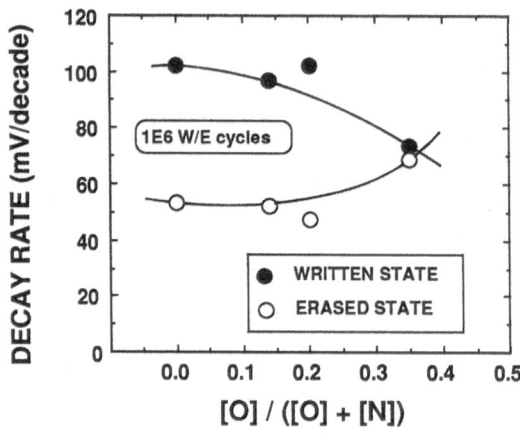

Figure 19. Decay rate of the stressed devices (10^6 cycles) containing an equal amount of stored charge, as a function of the [0]/([0]+[N]) value of the oxynitride layer.

device in figure 15, a number of conclusions can be drawn : first of all the decay rates are independent of the oxygen content for low x-values. The decay rates decrease for the written state for larger x-values as was also the case for the unstressed device. The decay rates of the erased state however increase for larger x-values in contrast with the results found on the unstressed devices. The decay rates of the written state are almost unchanged after cycling whereas the decay rates for the erased state are strongly increased after cycling and more or less follow the increase of the interface trap density. This would point to the following leakage mechanisms: the stored electrons (in the written state) are compensated by injection of holes [19], a process which is not influenced by the increased interface trap density; the decrease of trapped holes however is caused by backtunneling of holes, which is occurring through the interface traps and therefore strongly dependent on their density.

4. Conclusions

From this study the following conclusions can be drawn :
In the oxynitride layers a negative charge is built up which increases with increasing oxygen content. This charge is no longer compensated by a positive interface charge as is the case in

MNOS capacitors. This is believed to be caused by the high temperature hydrogen anneal carried out on these transistors. This net negative charge causes a positive shift of the center of window. For relatively low amounts of oxygen this effect is beneficial (zero center of window).

The programming of S-OxN-OS devices requires higher programming voltages for equal values of the memory window. This is partly due to the lower dielectric constant of oxynitride layers but also to the lower trap density in these layers with increasing oxygen content. Equal oxide fields in different S-OxN-OS devices do therefore not yield equal amounts of trapped charge.

The lower amount of trapped charge required for a fixed value of the threshold voltage results in lower threshold voltage decay. Equal amounts of stored charge (up to x=0.35) indeed yield the same threshold voltage decay rates in these devices, which in fact means that the charge decay for equal amounts of stored charge decreases with increasing oxygen content. This can be related to the lower trap density and the larger energy depth of the trap levels.

The degradation of S-OxN-OS devices is similar for the devices with low oxygen content. The best degradation resistance is however found for x=0.14. For large oxygen content degradation is considerably enhanced (enhanced decay rate in the erased state, increase of the center of window). The mechanisms for the charge decay for stored electrons and stored holes are identified.

In view of all these observations it can be concluded that S-OxN-OS structures with an oxynitride layer having a value for [O]/([O]+[N]) between 0.14 and 0.2 form better non-volatile memory devices. Their retention performance is significantly better and the endurance is slightly higher. The only drawback are the slightly higher programming voltages that are required. If no larger voltages can be allowed but a lower window can be accepted, this will result in an even better retention and endurance behaviour.

Acknowledgement

The author would like to thank J. Remmerie, E. Vandekerckhove, J. Witters and W. Beullens who were all at some time involved in the research activities on the use of oxynitride layers in non-volatile S-OxN-OS devices and contributed to the results described in this paper.

References

[1] *R.S. Bailey and V.J. Kapoor*, 6th NVSM Workshop, paper 5.1, Vail, 1983

[2] *H.E. Maes and E. Vandekerckhove*, Proc. Silicon Nitride and Silicon Dioxide Thin Insulating Films, The Electrochemical Society, Vol. 87-10, 93, 1987

[3] *P.C. Chen*, IEEE Trans. Electron Devices, ED-24, 584, 1977

[4] *Y. Yatsuda et al.*, Jap. J. Appl. Phys., 21, 85, 1982

[5] *E. Suzuki, H. Hiraishi, K. Ishi and Y. Hayaski*, IEEE Trans. Electron Devices, ED-30, 122, 1983

[6] *C.C. Chao and M.H. White*, Solid State Electron., 30, 307, 1987

[7] *A. Roy, F. Libsch and M.H. White*, Proc. Silicon Nitride and Silicon Dioxide Thin Insulating Films, The Electrochemical Society, Vol. 87-10, 38, 1987

[8] *T.A. Dellin and P.J. McWhorter*, Proc. Silicon Nitride and Silicon Dioxide Thin Insulating Films, The Electrochemical Society, Vol. 87-10, 3, 1987

[9] *J. Remmerie, H.E. Maes, J. Witters and W. Beullens*, Proc. Silicon Nitride and Silicon Dioxide Thin Insulating Films, The Electrochemical Society, Vol. 87-10, 93, 1987

[10] *J. Remmerie, H.E. Maes, M. Heyns, R. De Keersmaecker, F. Habraken, J. Oude-Elferink, W. vander Weg and A. Kuiper*, Proc. Silicon Nitride and Silicon Dioxide Thin Insulating Films, The Electrochemical Society, Vol. 87-10, 50, 1987

[11] *J. Remmerie and H.E. Maes*, Proc. Silicon Nitride and Silicon Dioxide Thin Insulating Films, The Electrochemical Society, Vol. 87-10, 189, 1987

[12] *H.E. Maes and S. Usmani*, J. Appl. Phys., 53, 7106, 1982

[13] *G. Groeseneken, H.E. Maes, N. Beltran and R. De Keersmaecker*, IEEE Trans. Electron Devices, ED-31, 42, 1984

[14] *J. Remmerie and H.E. Maes*, Proc. Insulating Films on Semiconductors, Eds. Simonne and Buxo, 15, 1986

[15] *H.E. Maes*, Ph.D thesis, K.U. Leuven, Belgium, 1974

[16] *H.E. Maes and G. Heyns*, J. Appl. Phys., 51, 2706, 1980

[17] *G. Schols and H.E. Maes*, Proc. Silicon Nitride Thin Insulating Films, The Electrochemical Society, Vol. 83-8 , 94, 1983

[18] *L.Lundkvist , I. Lundström and C. Svensson*, Solid State Electron., 16, 811, 1973

[19] *G. Heyns and H.E. Maes*, Appl. Surface Science, 30, 153, 1987

[20] *G. Heyns, Ph.D thesis*, K.U. Leuven, Belgium, 1986

CHAPTER 7

LPCVD SILICON OXYNITRIDES FOR LOCOS ISOLATION IN CMOS TECHNOLOGY

Jean-Luc LEDYS

MATRA MHS Semiconducteurs CP 3008 - Rte de Gachet -
44087 NANTES CEDEX 03 (FRANCE)

ABSTRACT :

A new isolation scheme, suitable for scaled down VLSI MOS circuits is
described, based on the use of LPCVD silicon oxynitride films as masking
material during selective oxidation. It is demonstrated that, varying the
overall composition of the layers, this material is definitely able to
circumvent the drawbacks of the usual LOCOS technique. Moreover, process
simplification occurs, in that scheme, which makes it a lot more attractive,
before any more advanced isolation technologies become manufacturable.
Finally, results are presented of the application of these material and
process to an industrial product : Thus, manufacturability is demonstrated.

1. INTRODUCTION

Chemically vapour deposited (CVD) silicon nitride layers, are, so far,
commonly utilized in semiconductors technologies ; low pressure CVD films
have been found to be suitable as oxidation mask in the so-called LOCOS
(Local Oxidation of Silicon) technique, capacitance insulators or storage
material in MNOS non volatile memories and as diffusion barriers. For back-
end applications, low temperature process capability is required : Thus
plasma enhanced or UV enhanced LPCVD layers are found as intermetallization
dielectrics or as final passivation cap. However, for most of the above-
mentioned applications, shortcomings exist : some of them have been overcome
by nitride-oxide sandwiches (passivation, gate or capacitance dielectrics),
but it recently appeared that single silicon oxynitride films could be
convenient either from applications or from processing point of view.

Indeed, LPCVD (or PECVD) films can be readily obtained in reactors similar to those already used for silicon nitride films, by adding N_2O to the usual mixture of NH_3 and SiH_2Cl_2 in the same temperature range. The overall layer composition can be simply varied to any value from SiO_2 to Si_3N_4 by adjusting N_2O/NH_3 gas flow ratio ([1], [2]). Doing that, it is expected that either the physical, chemical or electrical properties of these layers can be adjusted over a wide range, allowing new applications : while electrical properties' engineering of Si-O-N films may lead to improvements for dielectrics (intermetallization or MNOS structures), stress and physical properties' engineering allow to forecast interesting use as passivation layers or oxidation mask in a LOCOS sequence.

Since studies of film properties have already been extensively reported [3] [4], this paper will focus on the use of Si-O-N LPCVD films as oxidation barrier in the Localised Oxidation of Silicon (LOCOS process) used in MOS technologies for lateral isolation.

As the feature sizes are being shrunk, lateral device isolation is a major issue to consider in order to improve the packing density as well as the device behaviour in VLSI circuits. Classicaly, in MOS technologies, device isolation has been achieved by the so-called LOCOS technique. However, the scalability of the LOCOS process is severely limited by encroachment, including both physical and electrical phenomena, resulting either from lateral oxidation at the edge of the masked area for the former, or from lateral redistribution of the dopants coming out of the channel -stop ion implant, prior to the selective oxidation step, for the latter. Even if the second limiting factor, namely the electrical encroachment, is not likely to be only solved by a LOCOS-based isolation scheme, but by new device structures, we do consider that its effect is less stringent than the physical bird's beak effect. Hence, new schemes for isolation [5] [6] and for device structures [7] [8], recently appeared which suppress or decrease these effects and, on the same time, induce a smoother surface, more adequate for micronic technologies. Thereby, most of these approaches (i.e. trenches, semi or fully-recessed oxides) consider major changes in the process flows, as well as very new techniques and equipments, and so, are unlikely to be safely manufacturable in a short time. Thus, it is of greatest interest to find LOCOS improved isolation schemes, to sustain the near future MOS VLSI circuits technology.

2. EXPERIMENTAL

Entering the details of a usual LOCOS process flow, will allow to highlight the keypoints of such a technique (see fig.1) :

* To cope with the thermal mismatching between Si_3N_4 and bulk silicon, an oxide pedestal layer is grown underneath, which acts as a stress relief film. It avoids the growth of any possible crystal defects during the oxidation step, especially at the edges of Si_3N_4 mask, where stresses are maximum.

* The sandwich "Si_3N_4/Pad SiO_2" is used as a self-aligned implant mask (in case of using field transistors threshold voltage adjustment implantations).

* During the oxidation step (usually performed in the high growth rate range, to allow thick thermal oxide layers : 500 to 1000 nm), a tapered oxide edge is obtained, called "bird's beak", which stems from the lateral diffusion of oxidant species through the pedestal silicon oxide layer.

The stress engineering of LPCVD silicon oxynitride layers, by varying the composition (e.g. the O/N ratio) is thought to be able to overcome the bird's beak problem ; this paper is reporting on such a process.
So far, the suitability of Si-O-N layers has been checked, studying the behaviour of these films toward the main process requirements, namely :

- Patternability
- Oxidability
- Ion implant blocking efficiency
- Oxidation induced stacking faults generation

Parameters are the composition (O/N ratio) and thermal annealing treatment prior to any process step.
Silicon oxynitrides films have been deposited in standard LPCVD reactors (Philips Research Labs and MHS) using conditions already reported ([1], [2]). Patterning, ion implants and oxidation process have been performed at MHS using a 1.6 μm CMOS test circuit. Part of the oxidation study has been done in Philips Research Labs [4].

* INITIAL OXIDE
    To define the well implanted areas

* PR & IMPLANT & DIFFUSION OF WELLS
    To pattern and forms the wells

START P- TYPE
INITIAL OXIDE
N- MASK
N- IMPLANT

* NITRIDE DEPOSITION
    To be used as oxidation mask

* NITRIDE PATTERNING
    To define the active areas (non locos areas)
    => ETCH RATE & ETCH MODE

NITRIDE DEPOSITION
THIN OXIDE MASK

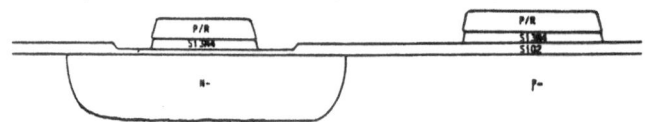

* FIELD THRESHOLD ADJUSTMENT IMPLANT (Boron 11 / 40 Kev)
    To raise the threshold voltage of field NMOS transistors
    => ION IMPLANT BLOCKING EFFICIENCY

P+ FIELD MASK
P+ FIELD IMPLANT

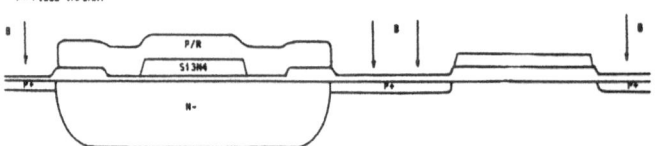

* LOCAL OXIDATION
    H2O oxidation at 950 C (H2 / O2 ratio = 2)
    To get 8000 A thick isolation oxide
    => OXIDATION RESISTANCE

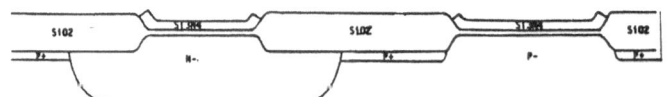

* NITRIDE / OXIDE STRIP
    To clear up the active areas prior gate oxidation

STRIP NITRIDE
CLEAN OXIDE

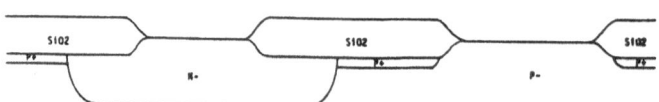

FIGURE 1 : LOCOS PROCESS FLOW DESCRIPTION

Composition and physico-chemical properties of the layers have been determined either by Rutherford Backscattering Analysis or by Nuclear Reaction Analysis (Utrecht University). Infrared Spectroscopy has also been used to assess the bonding configuration (by IMEC).

3. RESULTS AND DISCUSSION

3.1 Patternability

According to a conventional process, three kinds of etching characteristics have been assessed :

- dry etching for oxidation masked areas patterning : curves of fig. 2 and 3 give the results for two types of reactors in SF6 chemistry (reactive ion etcher at P = 1 x 10⁻² mbar and high pressure plasma etcher at P = 1 mbar). Decrease of etch rate versus the O/N ratio is more important in this second equipment. Annealing of films reduces the etch rate, for all the O/N ratios : this is thought to be related to either a structure modification (respective to H-Si-O-N bonds and concentrations) or to a densification effect [1].

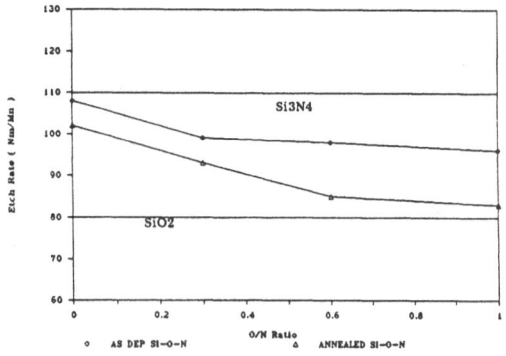

Figure # 2

RIE Etch Rate versus O/N Ratio

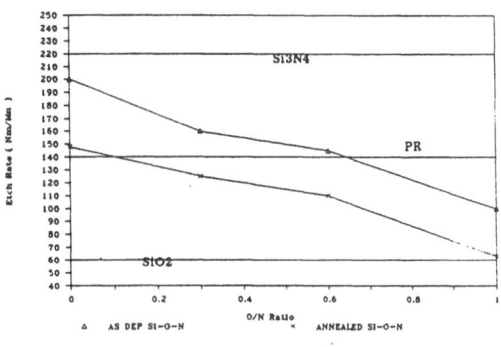

Figure # 3

HP Plasma Etch Rate versus O/N Ratio

- HF based etchants :

  Any cleaning sequence, involving fluorhydric acid, prior to oxidation must be highly selective in respect to the Si-O-N : for all compositions and O/N ratios, etch rate was found to be less that 100 A/mn in 11 : 1 BHF. No effect of annealing has been found.

- Hot phosphoric acid etching :

  After the local oxidation step, clearing up of the non-oxidized areas, involves a stripping of the masking layer. Usually this could be done by dry etching since underlying SiO_2 layer can act as an etch stop material, preventing any damage to occur at the active silicon surface. In our case hot H_3PO_4 has to be used, leading to a near-infinite selectivity over silicon or SiO_2. Etch rate was found to vary between 15 A/mn for O/N = 0 and 4 A/mn for O/N = 1. Annealing does not drastically change these values.

3.2 Ion implant blocking efficiency

More than a rather fundamental study, the margins around the usual process conditions have been investigated. Specifically this window was investigated for a N-well CMOS technology, where only a P+ field implant is required to increase the surface concentration of the P substrate underneath the isolation oxidized areas ; indeed the N-tub surface concentration is made high enough to ensure sufficiently large threshold voltages of parasitic field PMOS transistors. Tables 1 and 2 summarize the experiments and their results. In order to figure out this feature, surface doping concentration changes or type inversion due to ion implantation have been calculated from C(V) measurements, on N type silicon substrates.
Not surprisingly, the stopping efficiency of SiON is found to vary between that of SiO_2 and that of Si_3N_4. Again, a slight effect of anneal is shown ; the blocking is more efficient when the layer has been annealed prior to implantation. However, for O/N < 0.5, Si-O-N films thicker than 120 nm do withstand ion implant conditions, required by a single well CMOS technology.

O/N Ratio	Si-O-N Thickness	Boron II Energy (Kev)	Change in Surface Dopant Concentration
0	1000 A	25	Not Measurable
		35	-7.0 E15 At/Cm3
		45	N --> P (5.6E15 At/Cm3)
0.3	1150 A	25	Not Measurable
		35	Not Measurable
		45	N --> P (3.0E17 At/Cm3)
0.6	1120 A	25	-8.0E15 At/Cm3
		35	-6.0E15 At/Cm3
		45	N --> P (5.0E15 At/Cm3)
1.0	1170 A	25	-6.0E15 At/Cm3
		35	N --> P (2.8E16 At/Cm3)
		45	N --> P (8.0E18 At/Cm3)
1.0 (Annealed)	1170 A	25	Not Measured
		35	-4.0E15 At/Cm3
		45	-2.0E15 At/Cm3

TABLE # 1

Change in Silicon Surface Concentration after Boron 11 Ion Implant
(Implant Dose = 7.5E13 /Cm2 - Substrate Concentration = 4.0E16 At/Cm3 - N Type)

Si-O-N Thickness	Implanted Species	Energy (Kev)	Change in Surface Dopant Concentration
As Deposited 900 A	BF2	30	Not Measurable
		50	-1.0E14 At/Cm3
		70	Not Measured
	B 11	30	0.2E14 At/Cm3
		50	-0.9E14 At/Cm3
		70	N --> P (2.5E18 At/Cm3)
As Deposited 1400 A	BF2	30	Not Measurable
		50	Not Measurable
		70	Not Measurable
	B 11	30	Not Measurable
		50	Not Measurable
		70	-0.2E14 At/Cm3
As Deposited 2600 A	BF2	30	Not Measurable
		50	Not Measurable
		70	Not Measurable
	B 11	30	Not Measurable
		50	Not Measurable
		70	-1.4E14 At/Cm3

TABLE # 2

Change in Silicon Surface Concentration after 7.5E13 / Cm2 Dose Implantation
(N Type Silicon \ Initial Cs = 2.0E15 At/Cm3 \ Masked by As Deposited Si-O-N \ O/N = 0.3)

3.3 Oxidation

The study of the oxidation kinetics of Si O-N films has been described in chapter III.

In this work, films of thickness ranging between 80 and 150 nm, have been submitted to standard oxidation cycles (950°C/5H, 950°C/2H, 1050°C/2H in steam atmosphere), and have all been proved to be oxidation resistant for silicon oxide thickness ranging from 500 to 800 nm.

3.4 **Birds' beak formation**

After patterning in the RIE-SF6 reactor, Si-O-N films deposited on bare silicon have been submitted to the three oxidation cycles mentioned above birds's beak length (LBB) has been checked by SEM cross-sections. Fig.4 and 5 give the LBB dependence over O/N composition for 950 and 1050°C oxidation cycles.

These values have to be compared to standard LBB values in these two cases, namely 900 and 800 nm respectively.

The pictures the fig.7 show the modification of the shape of the bird's beak in the case of standard and Si-O-N Locos process.

3.5 **Oxidation induced stacking faults** (OISF)

The wafers described in 3.4, have been optically checked, for OISF density, after Secco decoration. Fig.6 shows how the defect density is affected by the composition and the annealing of the film prior to oxidation. We do notice here, that no true OISF, but only retrogrown OISF have been found during this inspection : indeed, our Locos oxidation recipes usually include an OISF retrogrowth thermal treatment.

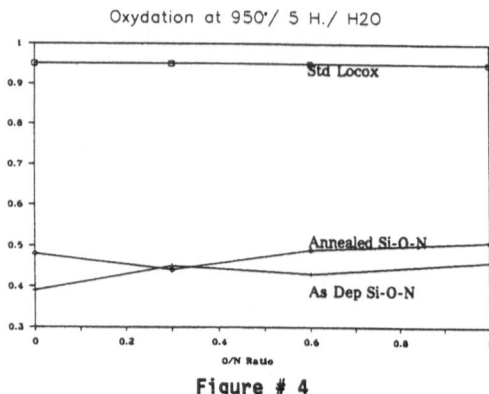

Figure # 4

Bird's Beak Length versus O/N Ratio

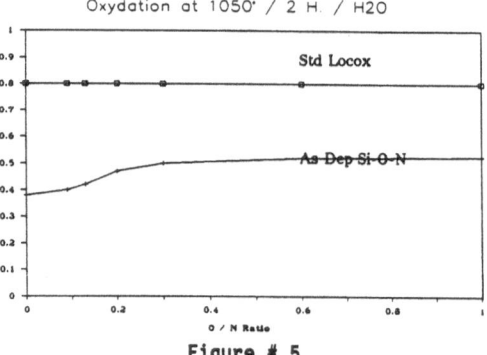

Figure # 5

Bird's Beak Length versus O/N Ratio

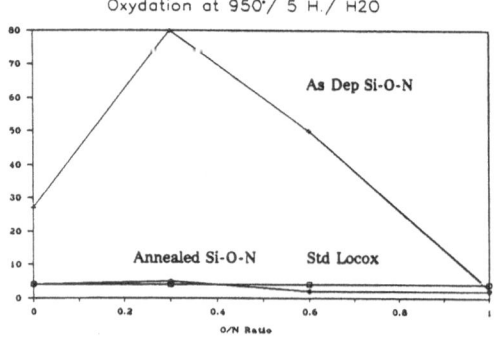

Figure # 6

OISF Density versus O/N Ratio

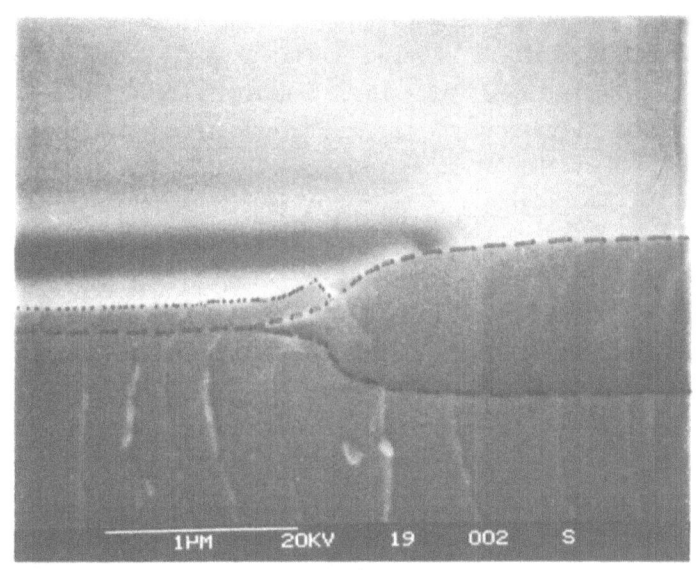

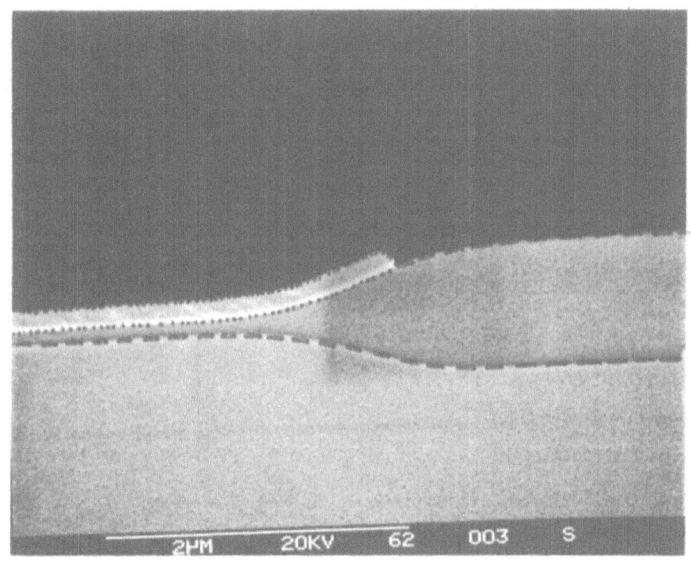

FIGURE 7

SEM PICTURES OF BIRD'S BEAK AT 950°C/5H
A : Si-O-N PROCESS
B : Standard Locox Process

Vacuum annealing at 920°C drastically improves these characteristics : If one reminds that [H] content is reduced after anneal, it could be expected from the hypothesis of § 3.3, that annealed films are more oxidation resistant. If so, volumetric changes during oxidation cycle are minimized, which would reduce the stresses and consequenthly the growth of crystal defects in the underlaying silicon substrate. The maximum OISF density at O/N = 0.3 could also be related to oxidation induced structural changes, since [H] content exhibits a maximum at this composition. Since OISF density is comparable to that obtained on reference samples, it could be reasonably thought that annealed Si-O-N is comparable to the sandwich $Si_3 N_4/Si O_2$ from the point of view of stress.

4. APPLICATION TO AN INDUSTRIAL CIRCUIT DEMONSTRATOR

4.1 Circuit demonstrator

The circuit which has been chosen to demonstrate the applicability of such a Locos process is a 16K bits fast access Static Ram Memory, produced since more than a year in MHS production line.

It utilises a full CMOS technology, featuring 1.5 microns minimum dimensions and micronic channel lengths (typically : 0.8 microns, either for P and N channel transistors). It also accommodates a 4 transistors memory cell with intrinsic polysilicon load resistors. The isolation is obtained through the Locos technique described in section 2.

4.2 Process

The first part of the process has been modified on one half of one lot, in order to allow the use of the oxynitride film instead of standard silicon dioxide/silicon nitride sandwich structure.

According to our previous results, the stoechiometry has been chosen to be 0.4 (O/N ratio).

The thickness of the film was frozen at 1500 Å for ion implant blocking efficiency. Finally an annealing treatment was given in order to improve the thermal mismatching between the film and the substrate, and thus to reduce the crystal defects density.

4.3 Results

--> **Die yield** as high as 80% of the standard one has been obtained, despite new process steps in particular the etching of the oxynitride layer.

--> **Effective width** (W_{eff})

The table below, gives for each type of transistor (N et P) the biasing of effective widths (Weff), ie the differences between drawn width and electrical effective width, as well as between electrical width and mask dimension.

parameter Type of transistor	W_{eff} - W_{drawn}	W_{mask} - W_{eff}
N Channel Experiment	- 0.48 microns	+ 0.92 microns
N Channel Standard	+ 0.18 microns	+ 1.58 microns
P Channel Experiment	- 0.34 microns	+ 1.06 microns
P Channel Standard	+ 0.13 microns	+ 1.52 microns

W_{eff} is improved by 0.5 to 0.6 microns. Bird's beak length is decreased by 0.25 microns/side.

--> Transistor threshold voltages

- Threshold voltages of P channel transistors are slightly lower (100mV) compared to reference values (surface doping level could explain that since hole mobility has been measured at 260 cm2/v sec instead of 230 cm2/v.sec)

- Threshold voltages of N Channel devices have been measured at 0.9V instead of 0.85 V, which is in accordance with P Channel variations.

--> Subthreshold current/fixed interface states density of active devices

In the table below are shown the subthreshold voltage slopes for N and P devices. These values are computed from $I_d = f(V_G)$ measurements performed for gate voltages $(V_G) \leq$ Threshold voltage (V_T).
From these subthreshold $I_d = f(V_G)$ curves, interface states density (N_{ss}) is calculated.

	subthreshold voltage slope		Extracted N_{ss}	
	reference	experiment	reference	experiment
N Transistor	80 mv/decade	80-90 mv/dec	1/3 E11 cm^{-2}	2/3.5 E11 cm^{-2}
P Transistor	120 mv/decade	100-125mv/dec	3/6 E+11cm^{-2}	4/7 E11 cm^{-2}

The behaviour of active devices appears to be not affected by the new process, leading to the conclusion that the Si-O-N Locos process, even if not very well tuned-up, does not degrade the Si-SiO$_2$ interface properties.

--> Field transistors behaviour

Threshold voltages are 2 to 3 volts lower than the reference values, but quite acceptable. This is thought to be due to bad adjustment of field threshold implant. Correction of such a problem is rather straightforward through adjustment of the implant dose and diffusion steps.

5. CONCLUSIONS

This paper emphasizes one of the possible applications of LPCVD silicon oxynitride films in VLSI CMOS technology : substitution in a Locos process of the usual Si$_3$N$_4$/SiO$_2$ oxidation mask by a single Si-O-N film, does drastically reduce the bird's beak length without any major additional drawback, compared to other improved Locos schemes (crystal's defects and dopants' diffusion due to high temperature process). This fits very well with the intermediate lateral isolation techniques, currently required for micronic MOS processes, taking place between the standard Locos and very advanced approaches.

Of course, in order to definitely comment on the usefulness of this scheme, compared to any other advanced Locos-based technique, a circuit demonstration has been undertaken : this has been done on a 16K SRAM circuit, using a 1.5 μm feature size CMOS process. Results pretty close to those currently obtained on this product have been achieved, although the process requires some additional minor tune up. Out of this first run, parts have been packaged and will be used for reliability studies : indeed, it has been noticed on such a technology, that the transistors drawn at minimum width and length, present an anormalous drift of their threshold voltages during burn-in stress test. It is expected that this phenomena is strongly size-related ; thus, those samples should exhibit a much more stable behaviour than the usual ones. This, if it is true, will add another dimension to the usefulness of such an isolation approach.

REFERENCES

[1] Physico-chemical and electrical characterization of LPCVD SiON layers F.HABRAKEN et Al. / ETW 1985 / ESPRIT project 369 . This book - Chapter I.

[2] A.E.T. KUIPER et Al., J.Vac Sci Technolog. 81, 62, 1983.

[3] F.HABRAKEN, A.E.T. KUIPER, MRS 1985, San Francisco

[4] A Study of the oxidation kinetics and mechanism is reported in Chapter III of this book.

[5] Kuang Y.CHIU et Al, IEEE Trans. on electr.devices, ed. 30,11, Nov.83

[6] H.P. VYAS et Al, IEEE trans. On electr. devices, ed.32, 5 May 1985

[7] L.C. PARILLO et Al, IEEE IEDM Conf. Washington 1980.

[8] Y.TAUR et Al, IEEE Trans. on electr. dev, ED 32, 2 Feb. 1985

[9] ENOMOTO et Al, Jap. J.Appl. Phys, 17, 1978

[10] CHRAMOVA et Al, Thin solid films, 78, 1981, 303.